Building Services Design Management

T0201436

Building Services Design Management

Jackie Portman

WILEY Blackwell

This edition first published 2014
© 2014 John Wiley & Sons, Ltd

Registered Office
John Wiley & Sons, Ltd, The Atrium, Southern Gate, Chichester, West Sussex, PO19 8SQ, United Kingdom.

Editorial Offices
9600 Garsington Road, Oxford, OX4 2DQ, United Kingdom.
The Atrium, Southern Gate, Chichester, West Sussex, PO19 8SQ, United Kingdom.

For details of our global editorial offices, for customer services and for information about how to apply for permission to reuse the copyright material in this book please see our website at www.wiley.com/wiley-blackwell.

Library of Congress Cataloging-in-Publication Data

Portman, Jackie.
  Building services design management / Jackie Portman.
    pages   cm
  Includes bibliographical references and index.
  ISBN 978-1-118-52812-9 (paperback)
1. Building–Superintendence.   2. Construction projects–Management.
3. Building–Planning.   4. Engineering design.   I. Title.
  TH438.P676 2014
  658.2–dc23

                                             2014012269

A catalogue record for this book is available from the British Library.

Wiley also publishes its books in a variety of electronic formats. Some content that appears in print may not be available in electronic books.

Cover design by Andrew Magee

Book illustrations: Dave Thomson

Cover images courtesy of
Fotolia_31720344 (**Air-conditioning Ducts – background image**) © **Phillip Minnis**
Fotolia_30377348 (**electric cables**) © **effe45**
Fotolia_38650039 (**Train**) © **jo**
Fotolia_43590673 (**air conditioning tubes**) © **AP**
Fotolia_51516109 (**air conditioning** ceiling) © **AP**
Fotolia_52506285 (ceiling with lighting) © **Photographee.eu**
Fotolia_56780047 (**front of building**) © **ginton**

Set in 10/12pt Palatino by SPi Publisher Services, Pondicherry, India

Printed and bound in Singapore by Markono Print Media Pte Ltd.

1   2014

Contents

Preface

Building services engineers seek to provide safe and comfortable environments for building occupants and for any activities happening within buildings: their remit may also extend to areas outside buildings. This process starts with the design of the appropriate systems and equipment, which then has to be installed and operated. It is all too easy for building services engineers just to concentrate on, for example, the water flow rates in pipes, the airflows in ducts, the temperature and airflow rate coming out of the diffusers, because these are specific. However, they also need to focus on what is happening within the space they are serving; for example, in any space there will be air movements due to draughts, leakages, window and door openings, and the buoyancy of the air will be changing from place to place; there may, or may not, be sun streaming through the window; heat is being given off by people, lights and equipment; ... and so the list goes on.

The role of a building services engineering design manager is becoming a discipline in its own right. There have been numerous efforts to place design on a higher intellectual level, and to develop design as a discipline with its own structure, methods and vocabulary. The methodologies for design management are inherently complex and the problem is exacerbated by the highly dynamic nature of the construction industry, the iterative nature of any creative process and the reworking that inevitably must be planned for. The increasing number of specialisms coupled with a tendency for participants to work in 'silos' provides further challenges. Finally, design management is increasingly becoming a contractor-led process which is a relatively new scenario for all the involved parties.

Traditional planning and management techniques are not well suited to the particular needs of the building services design manager. Design management issues cannot be resolved by squeezing the design process, achieving the same milestones with less information or making autarchic decisions to change design sequences. With respect to building services engineering, there are a lot of factors to be considered and many disciplines are involved. Non-existent or ineffective design management results in extended design timescales and poor quality of information. Any unresolved design issues have to be answered at some point in order for the installation work to happen. The effects of this can be increased costs, programme delays on site and inferior quality of the completed systems.

Design 'management' historically consisted of monitoring the drawing, document and schedule completion against a planned release schedule. This approach was crude and superficial, giving an approximate guide to progress without consideration of the design activity itself. The most serious inadequacy is the inability to predict the effects of changes. Design changes are an unavoidable outcome of the ill-defined nature of design problems. These arise frequently, owing to either the client's instruction – for example, a change/clarification of the brief – or the designer's eliminating an error or improving the design. Any technique that gives some insight into the impact of design changes (often termed 'design variations') on other design disciplines, the programme, on cost (to both client and designer) or on construction would be most valuable.

This book aims to give practical and relevant information to those involved with the design management of building services. In particular it recognises the idiosyncrasies and distinct features of building services engineering that are not specifically covered in general texts on design management – which tend to be architecturally focused. It does not provide specific guidance on how to design building services systems but it does contain direction on how to approach the management of the design.

The intended audience includes:

- building services engineering undergraduates, postgraduates and their tutors
- other construction-related discipline undergraduates and postgraduates and their tutors
- practising building services engineers who aspire to move into design management
- building services engineers who have found themselves promoted to design managers and need some support
- architecturally biased design managers seeking a better understanding of building services engineering design management
- project managers and clients in search of a better understanding of building services engineering.

'Purple panels' are included to offer some light relief from the main text. These provide worked examples, further explanations or useful background information.

While the book is biased towards the UK market in terms of references to terminology, legislation and working practices, the approaches are applicable to other regions.

The book is to all intent and purpose about management of a process. Yet successful design management, particularly building services engineering, needs leadership, which in turn means, good, even excellent interpersonal skills. These are about the how we communicate with, listen to, respond to, and understand others, such that problems are more accurately analysed and the corrective actions are more likely to remove the difficulty or resolve the problem, which contributes to the project's desired outcome and leads to our personally being more successful in professional and personal lives.

In a nutshell, this is the book I wish I had had during my career in industry, as I transitioned from a building services engineer to a design manager and as a part-time lecturer covering building services design management, when I would have appreciated relevant reference material to help structure my lectures.

About the author

Dr Jackie Portman DBEnv, MSc, BEng, ACGI, CEng, FCIBSE, MIET, MCIOB, MiMechE is a building services design engineer and manager with over 25 years' experience. She graduated in electrical engineering from Imperial College, University of London and took her first steps into the construction industry. She was attracted by the exciting, challenging, ever-changing and all-encompassing nature of the construction industry – where there are always new challenges and areas of interest – and she has never looked back. She has worked in consultancy, main-contracting, building services subcontracting, project management and client organisations in Europe, Africa, Asia and the Middle East.

She has led the design management process of a range of projects in terms of complexity, size and uses: university complexes (libraries, archive buildings, state-of-the-art education and research facilities), healthcare projects (wards, laboratories, clinical areas), single and mixed-use commercial office complexes, residential developments and schools. Her particular areas of expertise are in consultant selection and appointment, managing the design and pre-construction activities, and also in ensuring that commissioning management procedures are put in place, and closing out and handing over successful projects, and thereafter in instigating post-occupancy studies to understand how the building services engineering designs worked for the building occupants, operations and maintenance staff.

She fully appreciates the challenges of design management, where design issues cannot be resolved by squeezing the design process, achieving milestones but with less information or making explicit decisions to change design sequences. There are a lot of factors to be considered and many disciplines and stakeholders involved. Non-existent or ineffective design management results in extended design timescales.

She has always been keen to enthuse and motivate students and trainees and has used her 'hands-on' perspective to support full-time academics and teachers. She has been a visiting lecturer at the University of the West of England and the City of Bristol College, also contributing to the development of syllabuses, in particular, ensuring their relevance to current industry trends and requirements.

She obtained her doctorate from the University of the West of England, researching into ways and means of improving the contribution of building services engineers to the building design process, looking at how they are perceived by the rest of the construction industry and what tools and processes would help improve their performance.

Introduction

Building services engineering relates to the equipment and systems that contribute to controlling the internal environment so as to make it safe, usable and comfortable to occupy: this includes thermal, visual and acoustic comfort, as well as the indoor air quality. Building services are also provided to support the requirements of processes and business functions happening within buildings: manufacturing and assembly operations, leisure and entertainment facilities, medical procedures, warehousing and storage of materials, chemical processing, housing of livestock, plant cultivation and so on. For both people and processes, the ability of the building services engineering systems to continually perform properly, reliably, effectively and efficiently is of vital importance to the ongoing operational requirements of a building.

Building services engineers may be colloquially referred to as building engineers, architectural engineers, environmental engineers or mechanical and electrical (M&E) or, adding plumbing or public health, MEP engineers. The individuals and organisations involved as engineers are sometimes referred to as consultants or as designers. For consistency this book will use the term 'building services engineers' throughout.

Although contract terms and appointments do vary, building services engineers are generally responsible for the design, overseeing the installation and witnessing the testing and commissioning of some or all of the building services engineering systems. The design stages require technical skills to ensure that the systems are safe, compliant with legislative

Building Services Design Management, First Edition. Jackie Portman.
© 2014 John Wiley & Sons, Ltd. Published 2014 by John Wiley & Sons, Ltd.

requirements and good practices, are cost-effective and are coordinated with the needs of the other design and construction team professionals. The final design solutions may not necessarily represent absolute answers, but are rather the product of negotiation, agreement, compromise and satisfying the design criteria: as such, any assessment or measure of success can be relatively subjective and is difficult to measure.

With respect to the built environment, design management is the business side of design, which aims to create the right environment to control and support a culture of creativity and innovation, and to embrace the iterative nature of design involving the many disciplines that, collectively, will deliver design solutions – and all at the same time as ensuring that an organisation's commercial goals and objectives are achieved and that all is done in an ethically sound way. Typically the building services engineering installation is worth 30–60% of the total value of a contract, but existing literature on design management often bundles building services engineering up with other disciplines and does not recognise its unique features and idiosyncrasies.

Design management is not the same as project management. Project management focuses on a wider range of administrative skills and is not normally sympathetic to the peculiarities of delivering a fully coordinated functioning design, taking into account its unique nature and dealing with the changing requirements of clients and the external factors over which there is little control.

Evolvement of building services engineering

Throughout history the provision of air, light and water has been fundamental to the needs of people and processes. The fabric of roofs, floors, walls, windows and doors has altered relatively little over the past few centuries compared with the rapid and spectacular changes in the methods for servicing buildings. Figure 0.1 illustrates how some building services engineering systems have developed over time.

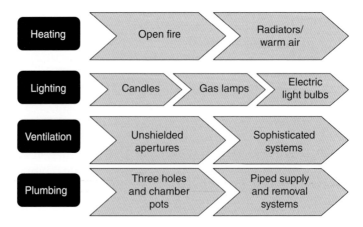

Figure 0.1 Developments in building services engineering systems.

The building services engineering content of buildings has grown at a phenomenal rate, decade by decade, both in quantity and complexity. From being a negligible proportion of the total costs of owning and running a building in the eighteenth century, the proportion absorbed by building services engineering has risen inexorably. What used to be the province of a craftsman now demands the services of a body of highly educated and specialist trained professional engineers.

Historically, building services engineering comprised mainly the heating, lighting, ventilation, plumbing and electrical distribution systems. With increases in the functionality and complexity of buildings, they now include much more: for example, intelligent building controls, medical and laboratory gases, communications and IT systems, transportation systems (for people and materials), security systems and fire detection and protection systems. Also, the correct provision of utility services to a building is fundamental, else it would be unusable.

The technology associated with building services engineering is changing faster than that in other parts of a building. There are two aspects of technology change to consider: first, developments in the building services engineering technologies themselves, these changes being mainly to do with improving efficiencies of systems, and second, new types of demand arising from technological change taking place in the activities of building occupants, the most evident case being computing and information technology.

Range of building services engineering systems in a building

The range of the building services engineering solution varies greatly in terms of the systems required, depending on whether a building is heavily or lightly serviced and whether the building services engineering services are intended to be covert or overt.

Some modern buildings are highly serviced and some less so, and the reasons for the high level of servicing will vary. Sometimes buildings, such as hospitals or laboratories, require a high level of building services engineering because they contain lots of specialised services, but others may require additional services because a building is located in a challenging environment, such as an area with high levels of noise or air pollution, both of which would preclude the use of opening windows for natural ventilation.

With 'low-energy building' there is an emphasis on reducing the quantity of building services by designing the facade and orientation of the building to minimise the heat gains and to utilise natural ventilation and daylight effectively to achieve comfortable conditions: the expectations of end-users may need to be persuaded not to expect perfectly consistent internal conditions. Whether a building has many or minimal building services engineering systems, it still needs to be designed and will still involve the skills of building services engineers in the analysis, design and management of the design – to ensure that the design actually works: for

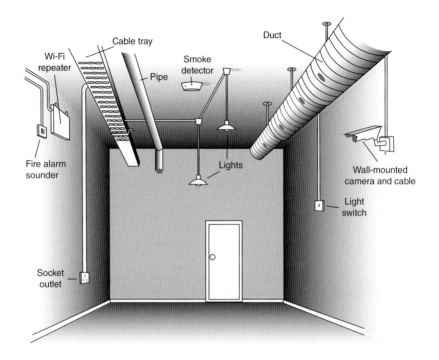

Figure 0.2 Overt building services engineering system solutions.

example, that you get adequate air movement through the building, that you can achieve acceptable lighting levels and that it is not too hot or too cold.

The plant rooms, which contain large pieces of equipment, are often located in basements or at roof levels, away from valuable usable floor space. Some building design philosophies may call for building services to be covert, with the likes of ductwork and cable trays an integral part of the interior design solution as illustrated in Figure 0.2. However, it is more usual for the majority of the building services engineering systems, particularly those associated with distribution to be hidden in the ceilings, walls and floors, with only the final termination units of any system protruding into the room – maybe a grille, switch socket, luminaire or a water tap.

Unique features of building services

The design process of building services engineering systems varies from the other main design disciplines (architectural, civil and structural engineering) for the following reasons:

■ *Dynamic nature of building services engineering systems* – Building services engineering systems have to react to changes in external conditions and in the patterns of behaviour inside a building – both of which are constantly changing. Nowadays the design process is aided by the use of computer models and simulations. These can simulate performance with respect to thermal behaviour, energy usage, lighting outputs (daylighting and artificial), ventilation and renewable energy sources, all with very many variables. However, even the best modelling programs cannot account for the unpredictable nature of the occupants and can only give a snapshot based on 'what-if' scenarios. It is very unlikely to be able to model every possible combination.

- *Occupant subjectivity* – Some aspects of the output of the building services engineering design are open to end-user scrutiny and personal judgements. Different people have different comfort levels and tolerances. What constitutes a comfortable thermal environment is a deceptively simple question but it has profound implications for building services engineers. Perception also varies according to tangible parameters such as age and gender. It is also tangled up with the psychosocial atmosphere at work and by job stress, making it difficult to satisfy all end-user requirements: even the perception of having 'control' contributes to a person being comfortable over a wider range of conditions such as the amount of daylight, heating and cooling. In addition, there are aesthetic considerations: some occupants will 'like' a particular luminaire/tap/radiator while others will not.

- *End-user behaviour* – Controlling the performance of building services is not just down to the installed building services engineering equipment and their controls. The design will be based on defined patterns of occupancy (density and duration of people in areas of the building, male/female/disabled ratios), assumptions with respect to portable equipment, for example, plugged into electrical sockets or connected to water outlet, and the nature of the finishes, including colour, density, texture and material, to walls, floors and ceiling. If the operation of a building deviates from the original design parameters, the design will be compromised and the building services engineering systems may not perform as expected.

- *Design life expectancy* – Building services engineering plant, equipment and systems are typically designed to function in a building for a lifespan of 20–25 years – at the most. However, in reality, this could be less, due to changes in legislation or as technological advancements which makes existing systems obsolete. This contrasts with the structural and civil engineering solutions which are usually designed for a much longer lifespan. Accordingly, the building services engineering design, at the outset, needs to take into account the likelihood that they will be upgraded or replaced in the future, and this includes considering how plant and equipment can be removed from the building, be responsibly disposed of and be replaced, while still taking into account of the ongoing operation of the building during the disruption.

- *Maintainability* – Building services engineering systems are the only active components in an otherwise passive shelter. The ability of the building services engineering systems to continue to perform interactively is of vital importance to the operational requirements. When a building is put into use, its building services engineering systems have to perform day-in, day-out for the life of a building and hence require ongoing attention. Shortfalls in design will be visible sooner or later.
- *Sequencing of the design process* – Despite the prevailing paradigms advocating multidisciplinary working, in reality architects and structural engineers still tend to lead the process of planning the building, so the building services engineering systems are expected to fit into the architectural and structural solutions.
- *Design responsibility* – Building services engineers usually produce drawings and a specification to obtain a tender. These should be coordinated with the architectural and structural engineering solutions. Notably, building services engineers do not produce construction or installation drawings. Their deliverables usually state the requirements passed on to subcontractors in terms of design responsibility. In contrast, architects and structural engineers usually produce drawings and specifications for contractors and subcontractors to build from, albeit, with details often supplied by specialists.
- *Energy consumption* – Building services engineering systems are a major consumer of energy. The current focus on sustainability and the green agenda means that more attention is being paid to energy consumption. This involves the operational efficiency of systems and selection of materials and managing end-user expectations of the systems.

Professionalisation of building services engineers

From the professionalisation of the construction industry emerged distinct subcultures defined by unique values, attitudes, languages, rituals, codes of conduct, expectations, norms and practices. These formed the basis of the rationales for professional institutions to be formed.

The bylaws and codes of ethics promulgated by professional institutions, including the disciplinary actions that will be taken against violations, are meant to reinforce assurance of proper performance of member professionals of the institutions. There are conflicting opinions as to whether or not the adoption of any such codes results in improved ethical conduct. Some commentators suggest that codes of ethics can never be more than 'window dressing' and, thus, self-serving as simply public relations efforts.

This book provides a wide ranging view of the environment in which building services engineers operates, the relationship with the rest of the construction industry including the other professionals involved, the technological content and the management skills which are needed to be able to manage the design process.

Part One explains the context in which building services engineering organisations operate, and how they relate to other stakeholders. This provides an insight into the dynamic forces pertinent to building services engineering entities (external and internal, local and remote) which drive their actions. Ethical issues are also addressed.

Part Two covers technical issues. It describes the particular tasks undertaken by building services engineers, how these relate to specific knowledge and skills and gives a flavour of the key challenges. This includes understanding design criteria, system descriptions and issues associated with designing for prefabrication.

Part Three focuses on the design process and management tools, the role of the design manager, interfaces of design management with other aspects of the overall project management, and the factors that will affect the future development of the design manager's role.

Part Four considers the particular design management issues of some specific building types.

In addition to the main text, 'purple panels' have been added. These aim to give some light relief, by means of anecdotes and examples, to help with enjoyment of the subject: the first sample describes some of the realities that building services engineers are faced with.

Case study: A reality check

Consider a typical office space: people sat at their desks and who would wish the air quality to be nice and comfortable at about head-height – say at about 1.2 metres above floor level. That's where the people are.

Unfortunately this is not where the building services are. They're round the edges, they're in the ceiling, they're in the walls etc. We cannot physically dangle bits of ductwork or a heater into that space very easily. So somehow we've got to try to achieve comfortable conditions in the middle of the room from systems that are all round the edge. Even worse, in order to achieve these comfortable conditions, we're going to need controls and sensors, so where do I place my sensors? Again, architects are not going to allow us to dangle them from the ceiling. If we have a nice space with some columns, I might think, 'Brilliant, I could fit some sensors to these at height level; I'll tuck them round the back of this column', but architects will again tend to object because it spoils the 'clean lines' of the space.

So now I'm stuck with having sensors and controllers around the edges of the space, and I'm trying to achieve conditions in a space which is some way away from where my systems are. So how do I know I'm achieving the right conditions? … The simple answer is that I don't! I can model, I can predict, I can look at heat interchanges in a space and be reasonably sure that if I supply air at these temperatures and if it's coming back at these conditions and I adjust it … then it's most likely to be acceptable in that occupied zone.

Now that's a horrible phrase, 'most likely' … but that's the distinction between the system hardware and the real engineering design.

Part One   The operating context

It is important to understand the context within which building services engineering design entities operate because this shapes and gives meaning to many things, and it can explain behaviours and actions. Context analysis examines the milieu in which an entity operates. There are many classification systems that are used, one of the simplest being to think of three different levels within the business environment – the organisational arrangement, the internal environment and the external environment to an entity – and how they are arranged for any particular project and the influences of the interfaces with stakeholders to a project. Ethical issues also need to be considered.

1   The operating environment

Organisational structure describes how an organisation is arranged: its hierarchy and how the components of this hierarchy work together to achieve the objectives set out in the vision and mission statements. This determines the operating procedures and the roles and responsibilities of the people employed.

Vision statements articulate organisational goals for the mid-term or long-term future. Ranging from one line to several paragraphs, a vision statement identifies what the organisation would like to achieve or accomplish. A good vision statement provides the inspiration for the daily operations of a business and shapes its strategic decisions. Vision statements may address a range of issues such as: aspirations for relationships with their clients, employees, project team members and suppliers; impact of their services in terms of sustainability and integration with the built environment: attitude to innovation: market positioning in terms of location: and types and sizes of projects. Some organisations may publish these in a very visible way, while others may keep them in-house or even keep them as verbal understandings.

Building Services Design Management, First Edition. Jackie Portman.
© 2014 John Wiley & Sons, Ltd. Published 2014 by John Wiley & Sons, Ltd.

Examples of vision statements

The following are vision statements, collected from the websites, of multidisciplinary organisations who provide building services design services:

Arup 'We shape a better world ... To enhance prosperity and the quality of
life ... To deliver real value ... To have the freedom to be creative and to learn.
Increasingly, we see ourselves as a provider of technical solutions; a design-
oriented technology house; a professional consultancy that has the stature to
exercise influence over those things we care most about. A fulfilment, perhaps,
of the most liberal interpretation of the original dream ...'

Atkins: 'Our vision – We aim to be the world's best infrastructure consultancy.
We define infrastructure as the systems that are vital for any nation's or
community's productivity and development, including transportation, utilities,
water, energy, large-scale built environments and information communications
and defence and security systems.'

Mott McDonald: 'To be the consultant of choice, recognised for the quality of
our people and our customer service.'

Mission statements are present-based statements designed to convey
how the vision statement will be achieved. They should inspire and give
direction to their employees rather than to those outside of the company.
As circumstances change, mission statements may need to be adjusted,
but they should always refer back to the same vision statement.

Examples of mission statements

The following examples are collected from a range of building services design
entities.
  We shall:

- provide superior client service and delight our clients
- act equitably and honourably with our suppliers
- protect and enhance the quality of the built and natural environment
- encourage innovation and creativity in design
- practice sustainable development.

Neither visions nor mission statements define how to achieve the
goals; however, by outlining the key objectives for an organisation, they
enable the organisation's employees to develop business strategies to
achieve the stated goals.

The business activity of any organisation can be classed into a particular economic sector according to the business types and the products produced: primary (extracting and processing raw materials), secondary (manufacturing finished goods), tertiary (providing services) and quaternary (providing intellectual activities). The provision of building services engineering design services is an example of the quaternary sector of the economy where the output is based on its intellectual capital. This is the possession of knowledge and experience, lore, ideals and innovation, professional knowledge and skill, good relationships and technological capacities which, when applied, will give organisations a competitive advantage.

1.1 Organisational arrangement

The organisational arrangement of an entity delivering building services engineering design may be configured in many different ways according to:

- ownership arrangement
- scope of building services engineering offered
- integration with other entities
- types of projects undertaken by building sector
- geographical operating span.

Ownership arrangement

Most organisations associated with building services engineering design entities are private sector owned, but some are in the public sector. In terms of legal structure, private sector organisations may be run as sole traders, partnerships or private or public limited companies. Public sector organisations are part of local or national government departments.

It is also possible for building services engineering entities to be under a licensing arrangement. Under this arrangement a licensor grants the licensee the right to use their name and, in return, the licensee submits to a series of conditions regarding the use of the licensor's intellectual property and agrees to make 'royalty' payments; for example, this arrangement may be used when, say, a UK based organisation wishes to open offices overseas and is prepared to have their name used under licence.

A building services engineering entity may link with other entities to form joint venture arrangements. The parties to the joint venture share (equally or otherwise) the provision of resources, risks and profits. The joint venture will have a unique name, which my use the names of the parties or it may be specific to the joint venture arrangement.

Whatever the ownership arrangement, building services engineering entities are usually independent of manufacturers, suppliers and installers of building services engineering equipment and systems. This enables them to offer unbiased opinions, judgements and decisions without any potential conflict of interest.

Scope of services

A building services engineering design entity may provide services for only one of the three main engineering disciplines that embrace building services engineering – mechanical, electrical or public health – or they may offer a combination of two disciplines, or provide all three disciplines.

Mechanical, electrical and public health engineering each comprise a range of core subject areas. With increases in the size of a project, the core subject areas are more likely to be dealt with by specialist building services engineers. With increasing complexity of a project more specialist subject areas will be added. Some entities may not be able to offer some or all of the additional specialisms for a particular project. Figure 1.1 illustrates this for lighting.

Some building services engineering entities may provide services for all stages of a project, from preparation to handover and operation in use, while others may specialise in certain stages only:, for example just doing feasibility studies and concept design work, or just doing the detailed design stage.

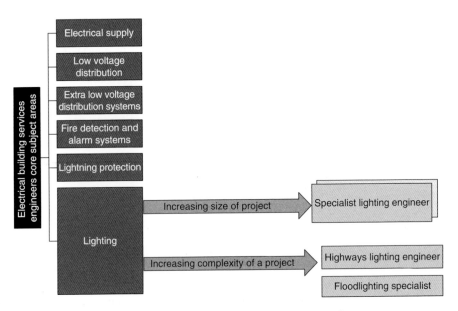

Figure 1.1 Effects of increasing project size and complexity on resources required.

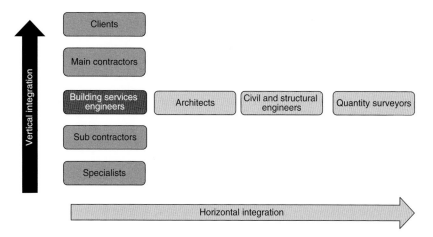

Figure 1.2 Vertical and horizontal integration of building services engineers.

Integration with other entities

The building services engineering entity may be standalone or integrated with other entities. This integration may be vertical or horizontal, as illustrated in Figure 1.2.

In a vertical integration arrangement, building services engineering entities may be part of a subcontractor or the main contractor organisation or a client organisation.

In a horizontal integration arrangement, building services engineering entities may be part of a multidisciplinary organisation with other design-related disciplines: civil and structural engineering, architectural or quantity surveying services.

These arrangements are not necessarily mutually exclusive, nor are they necessarily rigid; for example an in-house building services entity may also seek project work outside that organisational structure. These arrangements usually evolve over time, with entities being traded to fit into portfolios, usually with the ultimate aim of increasing efficiency and profitability.

Types of projects by building sector

Building projects can be classified according to building use as illustrated in Figure 1.3.

Some buildings services engineering entities will be able to provide services to all sectors, while others may specialise on a particular sector or subsector(s). Although building services engineers may be involved with buildings that support infrastructure projects, the design of the infrastructure for the utility services comprising distribution and transmission falls outside the remit of building services engineering.

Commercial buildings

Buildings supporting infrastructure

Industrial buildings

Complex multipurpose buildings

Residential buildings

Institutional buildings

Figure 1.3 Building sectors.

Geographical operating span

Building services engineering entities may be delivered from a single geographical location which may serve a specific local market, national market or international market. The services may be delivered from more than one office, again serving local, national or international markets.

1.2   The internal environment

The internal environment relates to the culture and climate of an organisation and will influence how resources are used to create value and attain goals. It includes factors over which an entity has some degree of control.

Capital, in the business context, refers to any asset that will produce future cash flows. The stock of capital determines the services

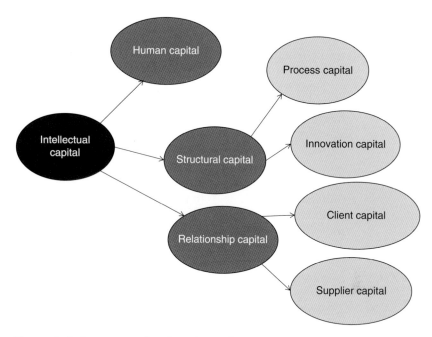

Figure 1.4 Components of intellectual capital.

delivered and their values. Intellectual capital broadly comprises human capital, relationship capital and structural capital. These can be broken down further as illustrated in Figure 1.4.

Human capital

A successful building services engineering entity achieves client satisfaction, provides technically sound professional services and maintains a supportive and rewarding working environment for its people. These aims are achieved by treating human resources as assets rather than expenses and liabilities. Human capital is a blend of knowledge, skill, innovativeness and the ability of each employee to fulfil the requirements of their job descriptions so as to add value to the organisation: when an employee leaves, the value of the human capital and their tacit knowledge also leaves.

Tacit knowledge is the knowledge that people carry in their minds and is, therefore, difficult to access. People may not be aware of the knowledge they possess or how it can be valuable to others. Tacit knowledge provides context for people, places, ideas and experiences. Effective transfer of tacit knowledge generally requires extensive personal contact and trust and falls under knowledge management.

To meet human capital requirements, entities must understand the core proficiencies they require and seek to employ and retain appropriate people. These competencies embrace a body of knowledge, supported by appropriate education and experience that result in effective job performance. These involve both 'hard' and 'soft' skills.

'Hard' skills consist of having the appropriate understanding of the concepts and terms in a particular subject area, and being able to translate ideas into practical designs expressed in written specifications, sketches, scale drawings and models.

'Soft' skill sets are equally important to the success of the building services engineering design to deal with the particular culture which building services engineers inhabit. This is influenced by the large number of people they have to work with during the lifetime of the project: these will be in-house and with external organisations. Individuals will usually have no choice over who they work with, and these relationships are subject to vulnerability owing to the temporariness of projects and the one-off nature of the product. Strong communication skills, both on paper and orally, are important to deal with the need to pass over information, liaise and discuss solutions with other parties.

In this transitional set-up, people have a tendency to base their relationships upon preconceived and standardised expectations of others' motives and behaviour patterns. The influence of stereotypes may also be strengthened by other factors such as the construction industry's confrontational, macho and time-pressured culture. Care is needed to ensure that any goodwill underlying interpersonal relationships, since it is delicate and precarious, is not destroyed by insensitive managerial practices.

If, in an organisation, the current pool of human resources needs supplementing, the main options are:

- recruiting suitably experienced permanent staff. This involves producing a job and person description, sourcing and selecting suitable applicants. Thereafter, a formal offer can be made
- hiring suitably experienced contract staff who are engaged either to provide a specific set of services or to work for a specific length of time. Contract staff may be either hired via an agency or as self-employed contractors. They are typically used to staff for peak periods in the work load or to provide specific expertise
- recruiting trainees, where the employer acknowledges that there are shortfalls in their competencies which they intend to address through providing on-the-job experience and training. Trainees are also permanent staff, but may have no experience or be part experienced. Raw trainees may be sourced from school leavers or graduates, or those seeking a career change. Part-experienced staff may be those already working in other building services engineering entities or who have transferrable skills, either

related to building services engineering such as electricians 'off-the-tools', technicians, facilities managers or those interested in changing discipline

■ subcontracting or outsourcing packages of work to a third party by entering a contractual agreement with an external entity to perform a certain amount of work that might otherwise be performed by in-house employees.

With the exception of subcontracting and outsourcing, where the responsibility for retaining staff is transferred to a third party, an entity's success in retaining staff, for the period they are required, is the result of the effectiveness of its human resource management efforts. This includes providing suitable mentoring and training, active career and professional development guidance, offering involvement in a variety of project types, creating a collaborative office environment and providing competitive salaries and benefits in order to motivate people.

As people develop, further skills are needed. These skills may include the ability to develop new business, manage time efficiently and communicate effectively with clients and team members. As the building services engineers move up the ladder in their entities they will undoubtedly become involved in planning and managing people and resources and directing activities ranging from billing and collecting fees to developing business plans.

Structural capital

Structural capital comprises process and innovation capital that support human and relationship capital so as to realise optimal business and intellectual performances in a repeatable and scalable way. When an employee leaves, the value of the structural capital stays in the organisation: it has organisational memory.

Process capital covers the value in creating infrastructure and business practices that support knowledge generation. The infrastructure comprises hardware, software, proprietary databases, drawing packages and information systems and the explicit information hosted on these platforms.

Hardware comprises equipment used to gather information, (e.g. cameras and video equipment), process it (e.g. computers and all their associated paraphernalia and networking equipment) and present it (e.g. projectors and screens, printers and plotters). There will also need to be means for data storage. Other hardware, such as smartphones and tablets contribute multiple functions.

Software may be used for calculations and modelling either for isolated or multiple elements of building services engineering design: for example, thermal performance, energy usage, lighting levels (daylighting and artificial), electrical distribution, ventilation patterns and impact of renewable energy sources. Software may also be used for checking and demonstrating compliance with codes and standards. Software is seen as adding benefits by being able to undertake calculations quicker and more accurately.

Databases are collections of information organised so that it can easily be accessed, managed and updated: for example, lessons learnt, standard details, standard reports and descriptions of previous projects and CVs of employees.

Drawing packages allow drawings to be represented on screen rather than having to be drafted by hand. This makes it easier to twist, stretch or move portions of a drawing, after which the information as a whole will automatically adjust. Users are able to zoom in and out for close-up and distant views. Depending on the particular package, it may be possible to switch between two-dimensional (2D) and three-dimensional (3D) views and to rotate images to view them from different perspectives.

The components of infrastructure need management to determine the requirements, specify and purchase the equipment, arrange delivery, put them in place, provide training, provide ongoing maintenance and ultimately dispose of the equipment. In the past, organisations typically developed their own software and databases in-house, some still do, but nowadays much proprietary software is developed, and databases created and managed, by specialist organisations and sold to organisations.

Information systems are a further development that combine hardware with business processes to support operations, management and decision making. They can generate models which may ultimately simulate the planning, design, construction and operation of a facility. The resulting model being a data-rich, intelligent and parametric digital representation of a building, including the building services engineering systems within. This allows views and data appropriate to users' needs to be extracted and analysed. This can generate information that can be used to make decisions and to improve the process of delivering the building.

Explicit knowledge is generally understood to be knowledge that has been or can be articulated, codified and stored so that it can be readily made available to others. The most common forms of explicit knowledge found in design are reports, specifications, procedures, drawings and calculations. These can also include videos, photographs, sound recordings and test results on samples.

Innovation capital is an intangible value driven by the recognition and mindset to solve a particular problem or the commercialisation of an invention. The outcome will comprise new products, services, processes or systems which may be legally protected as intellectual property and recognised by patents, copyrights and trademarks. Depending on their vision and objectives, organisations may make different strategic choices regarding their approach to innovation. There

Technology is not the total answer

Unfortunately, among many organisations and technology suppliers the concept of knowledge management has taken on a very narrow definition, to cover implementation of information technology to develop 'structural capital'. A common example of this is the misguided assumption that merely implementing shared databases or document libraries will enhance knowledge creation and use. While managing each element of intellectual capital is essential, it is seldom sufficient. Managing the integration of human, structural and relationship capital is the key to effectively building intellectual assets.

may be a range of approaches, from a fully resourced R&D facility staffed by building services engineers whose remit is to innovate, to encouraging a mindset of innovation as part of the day-to-day working of all staff. The extent of this support will be determined by the organisation's attitude to empowerment, creativity and risk taking. This may be one of the few instances when building services engineering entities relax their stance on being independent and autonomous of suppliers, and actively corroborate with them.

A balance needs to be struck between the risks to be accepted by building services engineers and the liability they are prepared to accept. As the liability of building services engineers increases, the innovation which they introduce may be diminished to the disadvantage of clients in terms of the project's life and/or life cycle costs.

Relationship capital

This consists of all value that building services engineers get from their clients, who pay for their services (either directly or indirectly as a subcontractor to an intermediate party), and suppliers who provide goods and services to them. Relationships are on both a personal and an organisational level.

The multidisciplinary nature of building design leads to very many relationships which are legally only limited by formal contractual lines. The lack of direct contractual relationships made the lines of authority subtle, hence the ability of groups to collaborate and work together, and this is, fundamentally, a function of trust. Trust is a valuable resource in the creation and use of knowledge. Formal contractual rules may bring about and legitimise behaviours and strategies that are at odds with common sense perceptions as to how trustworthy and cooperative exchange partners should act.

As building design complexity increases, the number of relationships involved rises, as does the level of involvedness of the relationships, all needing information from each other as a basis for their design decisions.

Depending on the contractual arrangement the 'client' to building services engineers may be the party procuring the building, a project management organisation, an architectural practice, a contractor or subcontractors – where they undertake the design work in the subcontractor's package. The relationship may be ad hoc, or more formal through preferred building services engineers lists or through framework agreements.

Building services engineers need to accept responsibility for making the relationships with clients work for mutual benefit and value gain. The value of relationships that an organisation builds with its clients leads to a more satisfactory working association, and this is reflected in their loyalty to an organisation and their services. This may be rewarded by repeat work, better fees, favourable references for other clients and recommendations for awards.

Suppliers to building services engineers provide the resources necessary to develop and deliver their services: for example, software for drawings and modelling and access to databases. Suppliers also include technical specialists who can assist by bringing new knowledge of their developments in service, or materials technology or other breakthroughs.

Building services engineers may work exclusively with one or more suppliers for a particular service. Supplier relationships may be ad hoc, or more formal though preferred suppliers' lists or through supply chain management procedures. Working with a variety of suppliers may insulate the organisation from potential setbacks if an exclusive supplier goes out of business or does not perform as required.

Summary

There are many different ways of organising how building services design is delivered. As an intellectual process, the entities providing building services design need their human, structural and relationship capital to be arranged to add value to the projects they work on. Individuals who are driven primarily by money are usually not the best fit for building services engineering design. Successful building services engineers tend to be motivated by performing at a high professional level for their clients, integrating with all the stakeholders in the process, delighting in problem solving and being rewarded both financially and in terms of professional recognition.

2 The external environment

The external environment to an entity comprises influences from outside its boundaries which could, in some way, influence or affect its ability to perform and deliver services. These can be divided into 'near' (meso-economic) and 'far' (macro-economic). The 'near' external environment includes competitors, while the 'far' external environment extends further afield.

Analysing the external environment involves searching for and acquiring information about drivers which may generate both risks and opportunities to which entities must react. They may be the root causes of the need for changes in the internal environment of an entity in order to secure or improve their position in the future. Some of these forces or drivers affect the construction industry players as a whole, while others have a more specific impact on the building services engineering entities.

The ability to respond to drivers in the external environment provides opportunities for service differentiation and improved competiveness through improved knowledge systems, reputation and credibility. This ultimately strengthens brand preference and improves business opportunities.

Building Services Design Management, First Edition. Jackie Portman.
© 2014 John Wiley & Sons, Ltd. Published 2014 by John Wiley & Sons, Ltd.

2.1   Competitor analysis

Competitor analysis involves entities, either formally or informally, benchmarking their performance against adversaries and assessing threats and opportunities from potential rivals. Information on competitors can be gathered from:

- *primary data* which is readily available: for example, annual reports and accounts, corporate magazines and promotional material
- *secondary data* which is the data that has been already obtained by a third party and collated into usable formats, for example, surveys, censuses and benchmarking information
- *opportunistic information*, which may or may not be actively sought out, may be 'anecdotal', coming from discussions with suppliers, customers and, perhaps, ex-employees of competitors.

Formal benchmarking is a methodical process of measuring and comparing parameters across organisations. It allows an entity to understand where they stand in comparison to their competitors and to identify industry best practices. This can be used to set improvement targets and to promote changes in the entity. There are many different parameters that can be benchmarked, some of which are very general to any business – for example, profitability, bid success rates, fee earnings, staff turnover, expenditure on R&D – while others may be more specific to building services engineering: for example, number of qualified staff in a specialist area. Before relying on any benchmarking information, entities should understand the basis on which the information was obtained and its limitations.

There are hundreds of benchmarking systems around the world, varying in scope, rigour and level of verification. The most stringent systems tend to be ones that are verified by an independent, third party organisation. The two most popular benchmarking systems related to sustainability are currently BREEAM and LEED. Others, such as the National Health Service's NEAT scheme, are specific to a particular group.

It is possible for an entity to undertake benchmarking itself, using information obtained from public domain record information. Alternatively, entities may collaborate with a group of others, in a combined benchmarking exercise by voluntarily contributing data for the benefit of all the participants. This allows the participants to choose entities of similar size and operations to make the result meaningful, and to ensure that an appropriate agreement is in place first to protect and maybe anonymise the data. Another option is to use information produced by an independent organisation experienced in benchmarking who gather information, possibly via surveys and interviews and provide an analysis of the results.

Benchmarking may provide some information with respect to possible opportunities and threats. Other information may be determined by analysing the market competition, evaluating client prospects and looking for the possibilities of new entrants to the market place.

2.2 PESTLE analysis

An environmental assessment will study and analyse the external environment to help understand the environmental influences on an entity, with the purpose of using this information to guide strategic decision making. The assumption is that if entities understand their current environment and assess potential changes they are better placed than their competitors to implement actions to respond to changes. There are many models available to assist with this analysis. This book uses the PESTLE model of environmental scanning to discuss the Political, Economic, Social, Technological, Legal and Environmental drivers in the external environment.

Political drivers

The arrangement of the political systems (the extent of democracy, power and influences) by governments, including the opposition parties, who are responsible for developing and shaping strategies and implementing policies impacts on the built environment. At an international level, relations between political cohorts may, in the worst case, lead to warfare and sanctions. Sanctions may impose a ban on trade, possibly limited to certain sectors such as materials and equipment and people. On a more positive note, international relations may lead to trade initiatives and the generation of international protocols.

Kyoto and Montreal Protocols

The Montreal Protocol is concerned with substances that deplete the ozone layer and it aims to protect the natural environment by phasing out the production of numerous substances believed to be responsible for its depletion. Among other things, this has affected the gases used in the design of fire protection systems and the choice of refrigeration fluids.

The Kyoto Protocol is a protocol aimed at fighting global warming by stabilising of greenhouse gas concentrations in the atmosphere at a level that would prevent dangerous anthropogenic interference with the climate system. This has led to greater consideration of how energy is sourced and used and any waste products disposed of within buildings.

In both cases, various countries have signed sets of international courtesy rules prior to endorsement by national governments, and these have been enforced by their national legislation.

Governments may commission reports into the construction industry which are intended to be used to inform policies and have made recommendations to the industry either as a whole or to a particular group such as clients, trade bodies or professional institutions. The outcome of these may directly impact building services engineering design in terms of how projects are procured and of the design standards and criteria used.

As significant spenders of public money on the built environment, governments are able to influence how projects are procured.

Economic drivers

The economic climate (local, national or international) affects the quantity and type of construction projects likely to be commissioned by clients. In a strong economy, with more finance available, favourable interest rates and greater demand for end products, more construction projects are commissioned. In turn, entities can exercise more flexibility in terms of fees demanded which, if successful, can affect expenditure on employees and equipment. In a weaker economy, fewer projects are commissioned and potential fee incomes are reduced, thus organisations may have less opportunity to invest in resources.

Entities located in one area but either undertaking work internationally or specifying equipment and materials internationally need to understand the effects of the second economy: exchange rates become relevant. This may affect decisions on staffing as employee costs vary in different economies. Also, the cost of certain materials may influence design decisions.

Social drivers

Social drivers are determined by demographics relating to the composition of the population and how groups within it are able to initiate social changes. As the population patterns change, the built environment will need to respond: for example, new schools for a growing young population. Also, a social movement, whose power then becomes harnessed and institutionalised in some sort of social code, often marks or provokes paradigm shifts which eventually become embedded in legislation.

Examples of social movements: Public health, sustainability and disability rights

Public health

The extensive drainage systems that remove foul water, which we now consider to be germane to our existence, were not always accepted as essential in urban areas. The expensive problem of public health infrastructure had to reach a critical mass before action was taken. Human waste is full of dangerous bacteria that can cause diseases such as cholera and typhoid. When this waste is not properly managed, it can come into contact with skin, water, insects and other animals that ultimately transfer the bacteria back into the human body where it can make people sick. As the problems of health increased there was a growing societal concern for public health engineering which led to action in the form of new infrastructure. Hence, legislation for public health emerged in response to a growing social driver.

Sustainability
Gradual appreciation that historical sources of energy, mainly based on fossil fuels, would eventually run out and alternatives were needed, coupled with awareness of the negative impacts of human growth and development on the built environment, has led to an increased consciousness for future generations. Although some legislation has been imposed and there are some financial incentives, there is a growing social momentum towards social responsibility by individuals and organisations. Within the past twenty years, the design, procurement and management of 'green' buildings has evolved from something of a fringe activity into the main stream.

Disability rights
Acknowledging that people with disabilities should have equal opportunities and equal rights as people was manifested in rights movements in order to make the issues public and to regain these people's rights. With respect to the built environment this applied to accessibility to and movement within buildings. Remembering that disabilities are not just confined to wheelchair users means that it's not all about providing ramps for wheelchair access but extends to elements of the building services engineering design such as lighting, access to switches and other controls.

Technical drivers

Three are two aspects to technical drivers: the development of new technologies which can be applied to building designs and the use of technology in the design process.

It is widely acknowledged that the technological infrastructure necessary to support virtual business operations is now readily available. Many writers advocate the virtues and benefits of using ICT as an aid to virtual collaboration and recognise the potential to improve visualisation of building design and construction. This technology supports construction project teams as virtual enterprises, which often comprise several disciplines, usually non-collocated, collaborating for relatively short periods in the design and construction of facilities. However, the root cause of building services engineering coordination problems may not be the lack of technology but tailoring its application to the specific technical and business conditions.

Be wary; technology has its limitations

Much of the recent work undertaken on collaborative working has focused on the delivery of technological solutions with a focus on web extranets, computer aided design (CAD), visualisation and knowledge management technologies, with seemingly very positive results and great potential – but there are practical obstacles:

- technical – ways need to be investigated to display digital media at full size on the construction site
- social – ensuring that all the stakeholder communities are involved; for example, local enforcing authorities and financiers, and small organisations such as specialist fabricators, may not yet be fully involved in the real estate strategies of healthcare institutions because of technical shortcomings
- legal – typically design team members caveat their information, and hence limit their liability, such that the site team cannot fully rely upon it. It is difficult to state that a model is not necessarily built to scale when models are supposed to accurately represent (and even behave like) a real building.

Legal drivers

Statutory requirements of many types have evolved to regulate the activities of those who wish to build; for example, planning legislation controls the appearance of buildings, building control legislation controls safety of finished buildings, and health and safety legislation controls safety of the processes of building. It is important to understand the hierarchy of legislation. Figure 2.1 illustrates the hierarchy in the UK.

Legislative changes can influence contractual relationships between different parties through the introduction of new case law precedents and new legislation.

With respect to non-statutory requirements, guidance material which may be included in the terms of a particular contract or drawn upon by virtue of good practice, may be sourced from client- or industry-specific standards, not-for-profit member-based associations, trade associations, all of which are designed for voluntary use or professional institutions. In the middle, there are British Standards (BS), designed for voluntary use and do not necessarily impose any statutory requirements. However,

Figure 2.1 Hierarchy of legislation in the UK.

AOPs, regulations, ADs and HSE publications may refer to certain British Standards and, thus, make compliance with them compulsory. Although, in law, most British Standards have only the standing of guidance documents, courts increasingly seem to place greater importance on them because of the consultative process under which they are drafted and the consensus within the industry that they reflect.

Environmental drivers

Building designs are affected by the water, weather and climate and eco-systems found in the natural environment which they occupy. Also, local materials, colours and textures may impact on the design of buildings. These may have an effect during both the life of the building and the construction period. Nowadays, most building designs aim to minimise their effect on the environment by addressing issues such as energy consumption (non-renewable energy depletion and fresh water consumption) and the disposal of waste products (greenhouse gas emissions, raw materials use and waste generation).

There is a growing awareness of the potential impacts of climate change, including designing for uncertain – and perhaps rapidly changing – climatic conditions, the need to adapt policies and behaviours and to develop possible mitigation strategies; these are affecting how building services engineers operate and the services they provide. This has led to a paradigmatic transition to a more energy-efficient building stock, through the adoption and implementation of a range of technologies and practices – and promotion and regulation of sustainability and environmental aspects – to enhance the environmental performance of buildings. This includes evaluations of the performance of buildings, and tools for documenting the environmental performance of materials and products. This has provided the building sector with important theoretical and practical lessons, and has encouraged better communication and interaction between the different actors within the building industry.

Summary

To help make decisions and to plan for future events, entities need to understand the wider 'meso-economic' and 'macro-economic' environments in which they operate. An entity is unlikely to be able to affect these factors, however, by understanding these environments, they have the possibility to maximise opportunities and minimise threats to the entity.

An understanding of competitors' past, present and, most notably, future strategies will help provide an informed basis to develop strategies to achieve competitive advantage in the future, and to help forecast the returns that may be made from future investments. Information on the past and present can be obtained through benchmarking, which then acts as a standard for judging the future against.

The business environment in which building services engineering entities operate is subject to PESTLE drivers over which they have limited control, but do affect the way they perform. Although some of these factors would equally affect other members of the design team, some were found to be more significant to building services engineers.

3 Engaging building services engineers

Clients will commission (appoint and pay for) the services of building services engineers to undertake work. Before any building services engineering design work starts, clients need to establish the nature of the services required and the criteria for selection of building services engineers to undertake them. This arrangement is legally formalised in terms of contractual obligations between the two parties, as well as coordinating with other parties.

In order for building services engineering entities to secure commissions they need to be able to promote and communicate the value of their services to potential clients. This includes marketing and business

Building Services Design Management, First Edition. Jackie Portman.
© 2014 John Wiley & Sons, Ltd. Published 2014 by John Wiley & Sons, Ltd.

development to position themselves for initial consideration, and subsequently the skills to respond to invitations.

3.1 Types of commissions

There are a variety of different commissions that a building services engineer may be contracted to undertake. They broadly comprise design, survey, advisory and witnessing – or a mix of these. They may also be appointed as contract administrators.

Design commissions

Building services engineers are usually commissioned as part of a design team. The team will be responsible jointly for carrying out the design and specification of the project and for the appointment of the contractor and his team. Design teams are also concerned with the installation works and quality control on site to the point of commissioning and handover, and thereafter with the occupation of the building. Overall cost control and expenditure management is a joint responsibility throughout.

The design may be for new-build on a greenfield or existing site, a complete or partial refurbishment of an existing building or base-build or fit-out works.

As far as building services engineers are concerned, in the area of major multidisciplinary projects, there are broadly three classes of design commissions: performance, abridged and full duties.

The procurement route will also affect the focus of the building services design. The broadest classifications are traditional (or Design, Bid and Build), Design and Build (D&B) and Private Finance Initiative (PFI).

■ The traditional route comprises a single-stage, fully designed project where the design is developed in detail by a design team working for the client, and a contractor is then appointed under a lump-sum construction contract which includes penalties for late completion. The contractor may have no responsibility for any design other than temporary works.

   Fully developing the design before tender gives the client certainty about the design quality and cost, but it can be slower than other forms of contracting, and as the contractor is appointed only once the design is complete, they are not able to help improve the buildability and packaging of proposals as they develop.
■ D&B involves procuring design and construction services from a single source, a D&B contractor, directly responsible to the client. The building services engineer's 'client' is effectively the main contractor. A principal feature is the ability to commence construction

before all of the design details have been completed. It places the responsibilities for both design and construction on the main contractor. One of the perceived benefits of the D&B method of construction is the ability to 'fast-track' projects.

D&B is suited to building services engineering design owing to its greater coordination and integration of the building team through partnering arrangement, and the avoidance of contractor's risk as a result of the nature of single point responsibility. This is probably also due to improved communication on the project and good working relationship among project participants

There will also be a different focus on costs because the building services engineer is designing for the main contractor's benefit.

■ PFI involves procuring design, construction and operation, for a defined period, from a single source, a consortium, directly responsible to the client. The building services engineer's 'client' is effectively the consortium.

The output of the design process, produced by building services engineers is published information covering:

■ showing how the building services engineering systems are coordinated with other systems – including allowing for space flexibility and divisibility, and to ease maintenance
■ proving compliance with regulatory requirements
■ allowing the installer to procure materials and equipment
■ allowing the systems to be safely installed, to comply with national safety standards
■ allowing the installed systems to be tested and commissioned.

Building services engineers may undertake some design commissions on which they act alone, rather than as members of a design team, but these are a minority, and are probably limited to the provision of single systems, such as provision of a standby generation or air-conditioning system, or refurbishment of an existing installation where the structure is not significantly affected, or fit-out works.

Survey commissions

Survey commissions will investigate existing situations and report on findings, without necessarily giving advice or recommendations. Areas covered are:

■ asset surveys which record the existence of equipment. This may also include their age, size and capacities
■ condition surveys which provide an assessment of the physical condition of the equipment and systems. The survey identifies potential deficiencies and maintenance issues

- dilapidations surveys which record the condition of equipment and systems and particularly comment on the extent of wear and tear
- energy surveys which determine how energy is used in a building
- occupant satisfaction surveys which record the views of occupants with respect to the indoor environment
- compliance surveys to determine whether systems and equipment are compliant with predetermined parameters, such as current regulations, future anticipated regulations or current thinking on best practice.

The level of detail of any survey will vary according to the information that needs to be reviewed and the methodology used for the study. Methods may include desks studies of record information, visual surveys of easily accessible areas, more intrusive surveys using basic tools and in-depth invasive surveys involving interfering with the steady-state conditions to obtain information.

A commission to undertake a survey may comprise a combination of different survey types. Surveys may be undertaken as a precursor to further studies; for example, an energy survey may be undertaken as part of a business case analysis to ascertain whether renewable energy sources should be introduced into a building.

Advisory commissions

Building services engineers may provide an advisory service which does not consist of detailed design or specification of works. This includes research, checking, reviewing, investigating and monitoring. Examples are:

- third party review of another building services engineer's design
- researching the latest technologies available for a particular application
- appraising the availability and location of public utilities services and their tariffs, to establish principles of supply: for example, high or low voltage electrical supply, medium or low pressure gas supply
- auditing existing and potential suppliers
- power quality analysis
- people movement studies.

Unfortunately disputes sometimes arise in respect of design matters or contractual situations. A special type of advisory role is that of alternative dispute resolution which is means of resolving disputes outside the court. There are a number of roles including:

- mediators
- adjudicators
- arbitrators.

In each role, building services engineers are required to be independent of the parties to the dispute and are expected to act fairly and impartially, using their knowledge of the subject matter in order to reach a fair decision, based on the evidence and arguments submitted by the parties. When appropriate, they take the initiative by ascertaining the facts and examining the legal aspects.

Mediators, adjudicators and arbitrators – what's the difference?

Mediators assist parties to reach a mutually satisfactory resolution and settlement of their dispute in a non-legally binding process. There is no resolution between the parties until all the parties agree to their own settlement. Without everyone involved saying 'yes', there is no deal and the settlement is non-binding until a settlement agreement is written and signed. Only after it is signed does it become a legally binding contract. A party can unilaterally withdraw from mediation.

Adjudicators also assist parties to come to resolution themselves. Adjudication is slightly more formal than mediation, but still without resorting to lengthy and expensive court procedures. The decision is final and binding, providing it is not challenged by subsequent arbitration or litigation. The parties are obliged to comply with the decision of the adjudicator, even if they intend to pursue court or arbitration proceedings.

Arbitrators listen to and consider the facts and evidence presented by the parties, applying the relevant laws and issuing a final legally binding award, but still without resorting to lengthy and expensive court procedures. Arbitration is a procedure in which a dispute is submitted, by agreement of the parties, to one or more arbitrators.

Mediation, adjudication and arbitration are all consensual and can only take place if both parties have agreed to it and to not involve attendance at a court with a decision made by a judge. If none of these options is followed, disputes between parties are determined by a formal legal route involving expert witnesses.

Expert witnesses can either act on behalf of either the insurers or the insured, establishing preparing, negotiating and settling insurance claims, or in respect of disputes between building owners/end-users and contractors or between contractors and subcontractors concerning technical or contractual aspects related to the building service engineering. All investigations and reporting are carried out on an independent basis. The principles employed are to establish the facts from site investigations, documents and other available primary evidence sources and to express an opinion based on that information.

Witnessing commissions

Witnessing is undertaken to ensure that the design criteria specified are fulfilled. These include partial or complete systems, and equipment. Witnessing may take place in the factory, in a dedicated testing facility or, for a completed installation, on site.

Witnessing in a factory, for items of equipment, will confirm that the requirements for the specification are met, particularly for custom fabricated units, such as air-handling units and electrical switchgear. This allows any modifications that are required to be made, in factory conditions, prior to transporting the equipment to site.

Contract administration

Alongside any of the types of commissions, building services may also be appointed by clients to act as their contract administrator. This role requires ensuring that all parties comply with the terms, conditions, rights and obligations of their contract, and also ensuring that the documentation for any tender is assembled. Furthermore, they must manage any changes to the agreement that might occur over the course of the contract and perform the closeout process when both parties have met their obligations. During all of this, communication is maintained with clients.

3.2   Contracts

A contract is an agreement, having a lawful objective, entered into voluntarily by two or more persons or a group of persons that comprise an entity which can be identified as one for the purposes of the law, each of whom intends to create one or more legal obligations between them. Contracts consist of offers, acceptances and consideration; they must be reasonable to both parties and legal, and not signed under duress on either side.

All projects will be governed by the terms of the particular contract. The form of contract communicates the procedures adopted in executing the project including the determination of the rights and obligations of the contracting parties. Building services engineers' contracts consist of a set of terms and conditions, setting out the obligations of the parties, including various schedules, for example, describing the scope of services, client's brief, programme of works, schedule of fees, deliverables schedule and particulars of the agreement – such as the names of the other involved parties. It may also include a memorandum of agreement (MOA) describing a cooperative relationship between two parties wishing to work together on a project or to meet an agreed-upon objective.

There are other relevant agreements: for example, those for the project managers, lead designers and all other designers and parties to the client. Ideally, these should be seen by everyone to ensure that all parties know what services others are commissioned to carry out. In addition, there are other more compelling contracts such as the project agreement and the funding agreement, and there may be others between clients and local authorities. All these agreements are between two parties, but they may have an impact on others who are not party to the agreement. They can contain specific requirements and conditions which impact on the building services engineers.

These agreements should be made available to all parties, but there may be some reluctance as they may contain sensitive commercial information. It can be done by redacting the sensitive clauses.

If they are not made available, building services engineers should record this omission and indicate the potential risks. Of course, if they don't even know of their existence then suitable wording is required to cover for the eventuality.

Standard templates for contracts are published by professional institutions and trade bodies. They may be amended by clients. Bespoke forms of contract may be developed by clients.

Allocation of design responsibility

The contract should determine the allocation of design responsibility. No literature has been able to unequivocally define design and hence the allocation of design responsibility is a contentious, and still unresolved, subject area: an unsettled debate. Organisations may have a vague or mistaken understanding of the roles of others with whom they must work; however, salient points are:

- in any design or management of the design, a party can only be responsible for that over which they have direct control
- the parties should make clear what has been included and the limits. The inclusion of an exclusion list should be considered carefully, as it is a list that can never be complete
- if building services engineers are not contractually required to complete aspects of the design, it follows that it must be made a contractual requirement in someone else's contract – otherwise there will be a gap.

Building services engineers and others should take an active part in discussions about how design activities can be clearly, fairly and openly allocated among the project team. The extent of design liability is not necessarily dependent on who does what, but rather on who accepts liability.

Big D and little d

The most important feature of design is that it involves envisaging and agreeing on the form of something which does not yet exist.

One idea to help is to use 'D' in Design to represent the concept, coordination and strategy of the works prior to construction and 'd' in design to represent the detail required during construction which takes it right down to 'how to screw a fitting to a wall'.

If this is supported by clear definitions of the drawing status and the precedence of the words in the specification over the drawings then the areas for misunderstanding are reduced.

'Reasonable skill and care' vs 'fitness for purpose'

In the UK, contractors are responsible for the provision of goods and the incorporation of these into the completed works, hence the obligation of a contractor in respect of the works falls under the Sale of Goods Act (1979). However, building services engineers are suppliers of a service (design) and thus fall under the Supply of Goods and Services Act (1982). With regard to the standard of performance, there is an implied term that the services will be performed with 'reasonable care and skill'. The most important practical problem encountered in a contract for a D&B project is where the subcontractors carry insurance liability to 'act with reasonable skill and care' but are not insured to achieve 'fitness for purpose'.

Provision of third party information

The contract should make reference to the provision of third party information. In the case of a refurbishment project, building services engineers should be provided with either as-built, record information or measured drawings. These all are intended to reflect the actual existing conditions but are subtly different.

- As-built drawings are prepared by contractors as they construct the project. These record the actual locations of the building components and changes to the original contract documents. These, or a copy of them, are typically passed to the architect or client at the completion of the project.
- Record drawings are prepared by architects when contracted to do so. These are usually a compendium of the original drawings, site changes known to the architect and information taken from the

contractor's as-built drawings. For some projects, this verification may require frequent or continuous presence on site.

■ Measured drawings are prepared from on-site measurements of an existing building or space. It can be for a building to which additions or alterations will be made, or for spaces which are intended for lease, and from which drawings the areas for lease purposes will be calculated.

All these are based on information provided by others. As the building services engineering equipment and systems are mainly concealed behind finishes, they can be difficult to verify. To do so, building services engineers would either have to observe construction continuously, full-time, throughout the construction phase, or perform intrusive and possibly even destructive investigations and testing after completion. The cost of such services to clients is usually prohibitive. Therefore, it is neither practical nor ethical for clients to attest to the accuracy of record documents or the accuracy of future design documents prepared on the basis of the information they contain.

To help minimise the likelihood of a misunderstanding or future claim concerning as-built, record or measured documents, it may be useful to include a clause in the contract that acknowledges the true nature of such drawings.

Warranties

Clients may also require professional appointments to provide rights to third party with an interest in a project. A collateral warranty is an agreement entered into by two separate parties which gives a third party rights guarantee in an existing contract. Figure 3.1 illustrates this principle.

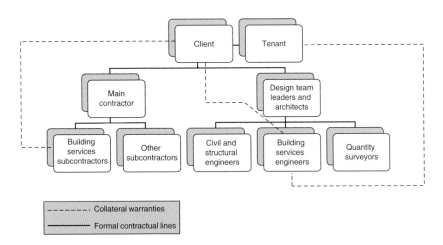

Figure 3.1 Collateral warranty agreements.

For example, in a situation where building services engineers have been appointed directly by architects, clients may wish to enter into a collateral warranty with building services engineers, for the benefit of their tenants. This imposes an extended duty of care and a broader liability on the two separate parties involved in a contract.

By creating a collateral warranty the parties are protected against an intermediate break in the contract chain.

Bonds

A construction bond is a three-party agreement between a surety, a principal and an obligee. Bonds are written agreements in which one party (the surety) guarantees that a second party (the principal) will fulfil its obligations to a third party (the obligee). If the principal defaults on its obligations, the surety must complete them or pay the completion costs to the obligee. The reason for it – for example, the principal's insolvency or their abandoning the work – is usually irrelevant. In exchange for guaranteeing the principal's performance, the surety charges the principal a fee called a premium.

There are many different types of bond. Bonds most relevant to building services engineers are performance bonds. A bond pays for pure economic loss, meaning the cost of completing the principal's obligation even if nothing is broken or destroyed.

Insurances

Building services engineers, along with the other members of a design team, owe a design responsibility to clients, providing advice, design, specifications, supervision etc., whether this be for a fee or gratuitously, and owe a duty of care to their client and to third parties.

The duty of care owed is generally the exercise of 'reasonable skill and care', in the discharge of the services provided. If building services engineers fail to exercise this duty and are negligent they may be liable for losses incurred by their client, and/or third parties; this may be, for example, by giving incorrect advice or making mistakes in calculations or by losing client information (including it being stolen or damaged). Claims against design building services engineers typically allege negligence, that is, a negligent act or failure to act by building services engineers in the performance of professional services. Taking into account the operation of the current legal system, even proving innocence can be very costly, and the insurance may include the cost of defence of a claim from a third party, so they are usually required by their clients to maintain professional indemnity insurance.

An insurance policy is a two-party agreement between an insurer and an insured. Professional indemnity insurance is applied at the level of the organisation rather than at the level of individuals. In the event of a claim, any potential claimant must prove negligence on the building services engineers' behalf, that is a failure to exercise reasonable skill and care in the discharge of their services.

If an allegation of negligence is upheld, the building services engineering entity is likely to be liable for the losses incurred by the claimant which arise as a reasonably foreseeable consequence of the professionals' actions. The building services engineering entity will often be responsible for the claimant's legal costs, and these can be substantial. It is often the case that large sums of money are spent simply trying to recover fairly minor losses.

Direct financial losses (economic and consequential loss), as opposed to the cost of rectifying a defect, is implicit under contract unless specifically excluded. There is more of a grey area arising out of the question of liability for economic and consequential loss at common law.

Partnering

Partnering is not a formal contractual arrangement but a way of working and a mindset which aims to increase cooperation across the organisation working on projects together. The results should be including improved communication among the project participants, productivity, lower costs, providing the product of construction to satisfactory standards and time, reduction in adversarial environments, supply-chain collaboration and more informed decision making for project participants. To succeed, partnering requires continuous and honest communication, trust, a 'win-win' attitude and a willingness to compromise, but an unwillingness to commit to the process can lead to ineffective construction partnering. Practical issues to be considered might include using information technology on common platforms – for example, BIM and podcasting – implementing innovative ways of working, such as using mobile and wireless communications, and collocating staff for critical periods.

3.3   Fees

There are several different methods of determining fees for building services engineers services:

- lump sum or fixed fee
- time basis
- percentage-based fee

A lump sum or fixed fee is an amount negotiated with the client for services that can be sufficiently defined at the outset of the project. This arrangement is best suited if the scope of the project, the schedule for design and approvals and the construction schedule and other variables can be determined with reasonable accuracy, thus allowing an accurate estimate of work hours and overhead costs to be determined. The fee then becomes effectively a fixed price, unless project parameters beyond the building services engineers control change. If these conditions change, or if the size of the project or scope of the building services increases or decreases, then the lump sum fee must be adjusted.

Time basis fees are charged at an agreed rate for a fixed time period, usually hourly or daily. This method of payment is useful when the services are difficult to determine in advance, owing to a vague brief, or if they are interim in nature and often short in duration. The actual rates vary according to the level of experience and seniority of employees and any subcontracted resources. The rates should be agreed at the outset, and a time period for review and adjustment of the rates should be set, in order to adjust for inflation and other factors.

A percentage-based fee links the fee for the building services engineer's payments to a percentage of the construction cost of the project. The percentage will primarily vary depending on the type of building, the construction value, the type of construction contract and the scope of services being delivered.

Generally speaking, percentage-based fees are based on sliding scales taking into account both the size and complexity of the project and the construction cost. The sliding scales are not suitable for many refurbishment projects or for very complex or custom projects. The fee indicated on the sliding scale is the starting point for discussion. It is a baseline fee which must then be revised using various fee adjustment factors to determine the appropriate fee for the building services engineering for the unique project. When calculating the distribution of the fee over the traditional five phases of a project, the following breakdown may be found on, for example, a 10,000 m² fully fitted-out office block, for full design duties.

Preparation	5%
Design	60%
Pre-construction	10%
Construction	20%
Handover and close-out	5%
Total Fee	**100**%

Typically, services are rendered and payments are made progressively, with final accounting for traditional basic services (100% of the total fee) at completion.

Often a combination of these various methods of compensation, rather than one single fee, are used; it may be appropriate to use one method of compensation for one phase of the project and a different method of compensation for another phase.

Example: Fee build-up

At the outset, when there is a requirement to visit the site and understand the existing installation and record information – which can be indeterminate in complexity and time – it may be fair to compensate building services engineers on an agreed to hourly rate.

As the details of the project become clearer and better defined it may be prudent to agree a fixed or percentage-based fee.

On larger projects that may last for several years, costs may fluctuate due to inflation affecting the costs of staff, accommodation and other resources, and due to changes in taxation. A cost, such as the building services engineer's fee, may be based on current prices, and then provision be made for them to be reimbursed for price changes over the duration of the project (a fluctuating price). This is may be based on pre-agreed published price indices.

3.4   Getting work

Building services engineers may either be approached directly to undertake work, respond to an invitation to undertake work or enter a competition.

The direct approach to undertake work may come from clients or contractors who are assembling a design team; or it may come from an architect or subcontractor who has been delegated the responsibility to select a team on behalf of clients. A relevant portfolio and informal references will assist an organisation in making it to the short list. The informal references may be from subconsultants and construction professionals, non-competitor colleagues, professional association members, not to mention friends and family members who have association with an organisation. This is enhanced by active business development. Also, on occasions there could be strategic reasons associated with business development for selecting a particular organisation. The whole process of formalising the terms of a contract may be negotiated.

An organisation may be working for one client and be novated, with the consent of all parties, to another with all obligations duties, with the terms being transferred. For example, this may happen if a client engages an organisation to undertake feasibility work, then, if a main contractor is engaged to undertake the construction work, the organisation is novated to the main contractor – primarily to look after the client's interests. The organisation will be subject to the terms and conditions of contract between the main contractor and themselves.

Clients and contractors may use frameworks. These are agreements with building services engineers to establish terms governing contracts that may be awarded during the life of the agreement. These set out terms and conditions for making specific purchases (call-offs). The use of such longstanding framework agreements with suppliers is professed to maintain continuity, reuse and transfer knowledge, reduce exposure to litigation and mediation, improve project outcomes in terms of cost, time and quality, increase opportunity for innovation and value engineering (VE), increase the chances of financial success and increase acceptance of identified risk and associated costs in the delivery of projects for a specific client. However, a framework agreement is more likely to not be a contract itself, but merely an agreement about the terms and conditions that would apply to any order placed during its life. In this case, a contract is made only when the order is placed, and each order is a separate contract.

Competitions may be organised by clients, ostensibly to allow any interested parties to put themselves forward for work. These may be 'open' competitions but more usually there will be constraints and rules, perhaps associated with size, portfolio and references. The ultimate prize may be an award of contract, with runners up being awarded cash. Before entering, potential competitors should ensure that competitions are genuine – do not assume fair play.

Responding to enquiries

Most enquires require potential building services engineers to respond to a pre-qualification questionnaire (PQQ).

A pre-qualification questionnaire typically includes requests for the following information:

- past financial statements, including costs and margins
- equal opportunity policies, practices and data
- health and safety policies, practices and data
- quality policies, practices and data
- environmental questionnaire, which asks about the environmental practice used, and copies of the environmental policy submitted for evaluation
- evidence of ability to deliver – track record of similar projects
- staff, including the number of chartered, non-chartered and technical staff
- portfolio of previous project
- references

The response to the PQQ is used in assessing the suitability of the organisation's commercial, technical and financial capabilities and provides a method of shortlisting interested parties meeting the required minimum qualification criteria. This aids the contracting authority in controlling the cost of the tendering process.

Thereafter a formal invitation to tender is issued to supply design services. This includes information describing the services required to enable the tenderer to prepare an accurate tender, which is in a prescribed format so that it is easy to compare with other tenders.

An invitation to tender might include:

- letter of invitation to tender
- form of tender
- details of any site visit, client presentation
- proposed form of contract, contract conditions and amendments
- tender pricing document
- any design-related information produced to date, such as architectural layouts or room data sheets
- instructions to tenderers explaining the tender process
- programme for the tender process
- methods for raising and feeding back queries during the tender process
- schedule of information required to be returned with the tender
- evaluation process and any evaluation criteria
- procedure in relation to alternative or non-compliant bids.

In response to an invitation to tender, tenderers submit their tenders, which will include their price for supplying their services. An evaluation and adjudication process takes place, and often supplementary questions are asked of the tenderers to clarify any issues. From this process a successful building services organisation is chosen to undertake the work.

Summary

Building services engineers are appointed by clients to undertake the work. This involves effort in terms of business development and marketing, together with the associated administrative effort. The details of the commission will depend on the client's requirements: this may involve design, survey, advisory, witnessing roles or contract administration. The arrangements will be formalised in a contract. Important points to remember are the allocation of design responsibilities, availability and reliance on third party information and the provision of bonds and insurances to protect other parties

4 Stakeholder interfaces

Building services engineers have to interface with a plethora of stakeholders who have a right and interest in imposing requirements to the end product. They are mostly unique, temporary, multidisciplinary, distributed teams set up for the duration of a particular project, and are often disbanded thereafter. Even when projects are procured using organisations from a framework agreement, the individuals involved may be different for different projects.

Figure 4.1 illustrates a stakeholder map, for different entities who may interact with building services engineers. These are broadly categorised as the client team, enforcing authorities, design team, utility services

Building Services Design Management, First Edition. Jackie Portman.
© 2014 John Wiley & Sons, Ltd. Published 2014 by John Wiley & Sons, Ltd.

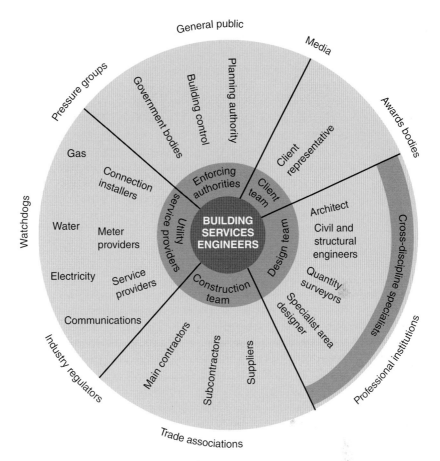

Figure 4.1 Potential stakeholder interfaces.

providers and the construction team, each of which has different objectives for a project. Within each of these categories, there are entities with primary responsibility for contributing to the overall objectives of the projects. These will be supported by secondary entities, such as sub-contractors and specialists. There are also other interested parties who have no contractual link to a project but who may have influences affecting the progress of the project.

Figure 4.1 shows the design and construction teams as separated entities. The concept of separating design and construction has been challenged in the past few decades. This has led to new practices, new roles and new working relationships among the different disciplines of building professionals, and the roles of building professionals in the structure of a project team have been changed. In reality building services engineers (and the other members of the design team) are usually discharging some of their duties under the construction team heading.

The particular stakeholder make-up and the contractual arrangement between the stakeholders varies according to the particular project.

In some cases, the function will not be required or more than one function may be undertaken by the same person; for example, the architect may also be the landscape architect.

The diversity of stakeholders makes for a multidisciplinary culture and complex issues with multiple and overlapping influences which are legally only limited by formal contractual lines. This lack of direct contractual relationships makes the lines of authority subtle and may be a factor in creating a less than optimum team performance: formal contractual rules may bring about and legitimise behaviours and strategies at odds with common sense perceptions as to how trustworthy and cooperative exchange stakeholders should act.

4.1   The client team

Clients are responsible for initiating, commissioning and paying for the design and construction, and ultimate disposal of the facility of a building. This includes providing the land and finance, defining the scope of the project and determining key programme milestones. They are also responsible for communicating to the design team their needs and objectives by initiating the project briefing process. Effective briefing requires clear definition of the client's requirements and communication to the other stakeholders. The emphasis should be on the objectives of the client in terms of building requirements, cost and time budgets rather than on the provision of solutions.

The basis of any project should be determined by a business case analysis. This allows those managing resources to analyse the rationale, by assessing the impact of the proposed options and comparing these against other factors, such as the major risks and any other drivers in the organisation, in order to decide which, if any, of the options would be worth their while pursuing. Consideration should also be given to the option of doing nothing, including the costs and risks of inactivity.

Client teams consist of individuals, internal departments and external organisations organised to deliver the client's responsibilities. Project sponsors are the individuals (often a manager or executive) who act as the representative of the organisation with overall accountability for the project. They are primarily concerned with ensuring that the project delivers the benefits set out in the business case and should be mindful of, and react to, any circumstances which necessitate a change to the business case. They need to establish a structure and protocols for decision making. They may specify that design team members will be benchmarked, especially for documentation quality produced, so as to provide a mechanism for monitoring their performance, and possibly to motivate participants by establishing realistic goals demonstrated to be achievable.

Client teams should provide a single point of contact, usually a client representative, to interface with the design and construction team, to provide information and decisions on behalf of the client team. The main roles of client representatives are to have a full understanding of the project, and to be able to communicate this. Client representatives coordinate end-user inputs, assist with preparing the project brief, control changes and risk, manage the project budget and programme, secure professional services as required, determine the procurement route, manage reporting arrangements and provide a focal point for contacts within client teams.

Clients will appoint contract administrators to act on their behalf for the purpose of administering the construction team's contract. The contract administrator may already be appointed as part of the design team – for example the architect, quantity surveyor or building services engineer – or may be an independent party.

Project sponsors and client representatives are supported by a number of other specialists. These competencies may either be in-house or out-sourced.

- Health and safety – this includes a duty of care for maintaining, and when required providing, accurate record information of the building services engineering installations, including the currently installed private utility services within their premises, associated with their building's information. It also includes determining the way in which the fire management of the building will operate in an emergency.
- Legal – this includes responsibility for drafting project specific contracts, including any bonds, purchasing land and involvement with any contractual dispute issues.
- Financial/commercial – this involves ensuring that finance is available and distributed in accordance with the contract.
- Facilities management comprising building operation and maintenance.

   Building operators may also provide furniture, fixtures and equipment (FF&E). These items vary in nature but typically include IT/telephony, audiovisual equipment, signage, specialist equipment and catering appliances to support facility functions. Some items may be fixed, while others are movable. These need to be considered as part of the design and construction, particularly with respect to requirements for utility supplies and the heat gains to the space.

   Building operators are responsible for ensuring that buildings are used as per the design parameters applied to the building; otherwise they will be in danger of compromising them and the building services systems will not operate as intended.

   With respect to maintenance of building services engineering systems, the client team needs to advise on the plant operating pattern, availability and product quality within the accepted plant conditions (for longevity) and safety standards, and at minimum resource cost. They also need to contribute their knowledge and requirements with respect to access to equipment for cleaning, maintenance and removal, and also space required for personnel and their functions, such as office areas, stores for spares, workshop areas and loading areas.

■ Letting agents are responsible for marketing and agreeing terms and conditions for tenants. With respect to the building services engineering this means ensuring that relevant design criteria – for example, availability of socket outlets at desks – are clearly stated and, if the provision is not acceptable, to be able to discuss.

Depending on the size of the project, client teams may include clerks of works as their 'eyes and ears' on-site in order to make decisions on behalf of the client, to monitor progress and the level of resources being deployed and to ensure that the required levels of workmanship are achieved.

4.2   Enforcing authorities

Statutory requirements of many types have evolved to regulate the activities of those who wish to build: planning legislation controls the appearance of buildings; building control legislation controls safety of finished buildings; and health and safety legislation controls the safety of the processes of building. There may be legitimate ways and means of deviating from the particular requirements of the enforcing authorities.

Legislation in the UK

Requirements for buildings are embodied in (a) Acts of Parliament (AOP); (b) regulations, made under the powers confirmed by an AOP and typically embodied in Approved Documents (AD), which provide guidance about compliance with specific aspects of the regulations and what is likely to be accepted as a reasonable provision for compliance, but can be overturned if the case is in some way unusual; (c) Health and Safety Executive (HSE) requirements which provides interpretation of the AOPs to help with compliance and to give technical advice. A detailed schedule of the statutory controls currently in force, selected Acts, regulations and compliance information in the UK relevant to building services engineering can be found at the Building Services Research and Information Association's (BSRIA) Legislation and Compliance Online Database (BSRIA, 2013).

Building control

Health and safety is a major reason for governments to produce regulations for the built environment. In the course of time other issues, such as energy management, special needs, sustainability and economic motives have been included. For these subjects technical requirements are formulated and the procedures for checking building plans against the requirements and issuing the building permits have been laid down in law.

Local planning departments

If the project consists of a new build, a major refurbishment or a change of use, then it is likely that planning permission from the local planning department will be required. The process of obtaining planning permission is usually led by architects with building services engineers contributing in the following areas:

- Input to the orientation of the building and the elevations with respect to size and locations, or louvres and windows with respect to heating and cooling loads and the availability of daylight – also the location of any external, visible building services equipment.
- Developing a utilities statement describing how a building will connect to existing utility infrastructure systems – building services engineers will demonstrate that the availability of utility services has been examined in consultation with the relevant utility companies and the proposals would not result in undue stress on the delivery of those services to the wider community, that the design incorporates any utility company requirements for substations, telecommunications equipment or similar structures, and that service routes have been planned to avoid the potential for damage to trees and archaeological remains as far as possible. With respect to foul and surface drainage, if the development does not involve connection to foul or surface water sewers the alternative means must be described. If soakaways are proposed, percolation tests must be carried out and the assessment should include details of the results.
- Developing a ventilation/extraction statement for buildings such as restaurants or cafes, or where the process means that fumes have to be extracted – building services engineers will need to produce plans showing the location of the equipment, including the location of external vents, detailed drawings of the design of the equipment, pipework and flues, including full details of its external appearance, where appropriate, any odour abatement techniques employed and the acoustic characteristics of the equipment and measures proposed to minimise the impact of noise on neighbours (including noise insulation and hours of operation).
- Providing information pertaining to any artificial lighting – the details may include a layout plan showing the location of all light fixtures and beam orientation and spread patterns of illuminated areas with specified lux levels, elevation details showing the position of the lighting units (whether freestanding or attached to existing buildings or structures, a detailed performance specification of the equipment proposed, the proposed times at which the lighting

will be in use, an assessment of the impact of the lighting on the adjoining uses and the locality generally, and mitigation measures to remove or reduce any adverse impacts identified.
- Discussing the impact on listed building or building of architectural or historical interest.

In addition, a local planning authorities may consult with statutory consultees. These are public bodies that need to be consulted on specific aspects of planning applications.

Non-departmental public bodies

Non-departmental public bodies can bring a degree of independence and offer expertise to deliver public services. Some are only able to give advice, while others have regulatory power, for example with respect to health and safety.

4.3   The design team

The design team consists of the various disciplines required to develop and deliver appropriate solutions to meet the client team's requirements, and to prepare the required level of information for the construction team. Collectively, the design team is responsible for translating the client's requirements into information that describes the scale and shape of the proposed building as well as the materials to be used. The level of information produced will depend upon the particular project.

The main design disciplines are usually seen as architectural and engineering, but quantity surveying (the management of costs and contracts) should be included. There may be any number of other specialists involved. These may either by in-discipline specialists who have more detailed knowledge within a particular discipline or cross-discipline specialists who provide broader knowledge across disciplines, or specialist-area consultants who have particular knowledge associated with the end-use of a building.

Architects

Architects are primarily concerned with the form and appearance of a building – on the inside and the outside – to suit the way in which people and processes make use of the building, as well as fulfilling an aesthetic agenda. To achieve this they will need to seek advice from engineers.

At the outset, architects are responsible for interfacing with the client to develop the design brief: this will allow for expectations, project requirements and budget. They will use this information together with the building site conditions to determine the best location and orientation, taking into account requirements of enforcing authorities, on issues such as planning and building regulation approval and life safety matters. This will allow the development of conceptual plans, sketches and models. These are further developed into detailed drawings and specifications for use by the construction team.

During the construction phase, architects work with the construction team to ensure that the project is constructed in accordance with the drawings and specification. After they are constructed projects have a warranty period – the defects liability period. Architects may be responsible for following up any relevant issues or outstanding work with the client and the construction team.

Typically architects are in the lead position in the design team, which usually includes the lead design coordinator role, although, in reality, the design coordinator cannot be an expert on all information needs of every design discipline and will not necessarily have a realistic overview of the entire project.

Building services equipment and systems need to fit into the building design. Sometimes they will need to be hidden from end-users, with rooms, routes and risers being provided to accommodate building services equipment, while other aspects will be very visible, such as terminal units.

Architects may be supported by specialists including:

■ master planners, who are concerned with a conceptual layout for a site considering existing buildings, the next phase of growth and the future growth
■ landscape architects, who are concerned with safety, aesthetic, environmental and operational issues of outdoor spaces, including hard (built) and soft (planted) materials. The landscape scheme may need to accommodate lighting, small power, security, ICT requirements. It will need to take account of access to and constraints (wayleaves and easements) associated with utility services located in the area. Also, the water balance between the land and the building needs to be coordinated: surface water can be collected and used within a building, wastewater from a building can be used for irrigation. Further sub-specialists are:
  ○ arboriculturists who are more specifically responsible for carrying out the tree assessments, specifying tree protection measures, monitoring and certification. Typically interfaces between building services engineers and arboriculturists will be ensuring utility services distribution lines and apparatus are coordinated with the tree roots and foliage. There is a variety of minimum distances that must be kept from utility services as illustrated in Figure 4.2.
  ○ ecologists study the relationships between plants, animals and their environments; architects need to appreciate where their building design may impact on any of these, and whether any mitigation methods are required

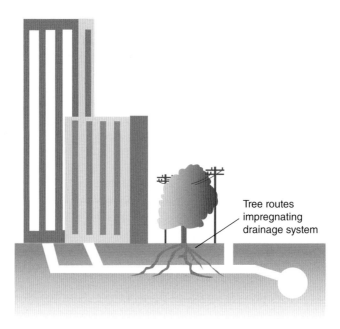

Tree routes
impregnating
drainage system

Figure 4.2 Trees impacting on drainage pipes and overhead electricity cables.

○ archaeologists are interested in any disturbance of land within any area that is identified as having archaeological importance (such as listed buildings or structures, historic parks and gardens and historic battlefields) or is a designated as an ancient monument, or has been otherwise identified as having potential archaeological interest. Archaeologists may either undertake an archaeological survey or provide a watching brief where the development is in an archaeologically sensitive area. The relocation or provision of new utility services may involve earthworks which could potentially affect archaeological heritage which will necessitate mitigation measures to remove or reduce adverse impacts

■ interior designers are concerned with creating a physically comfortable and aesthetically pleasing interior environment for the occupants, as well as the space links to business strategies and goals. Terminal units of building services equipment – for example, grilles, diffusers, socket outlets and switches – will need to be coordinated in terms of appearance and location. Access to the building services engineering systems distribution system may require the provision for openings (access hatches, lift-off panels and doors) in the finishes. The colour schemes and finishes will affect the lighting design: darker colours absorb more light than lighter colours hence more light is needed to achieve the same lighting levels.

■ facade engineers are concerned with the frontage of the building. As well as the overall desired aesthetic for the facade this includes the airtightness, thermal performance (heat losses and solar gains), daylight penetration and fixings and cableways for luminaires, CCTV cameras and other equipment.

Engineers

Engineers are concerned with the elements of building design that support the architecture. The main engineering disciplines are building services engineering and civil and structural engineering, both supported by in-discipline specialists.

■ Building services engineers are concerned with controlling the internal environment so as to make it safe, usable and comfortable to occupy. This includes thermal, visual and acoustic comfort, as well as the indoor air quality to support the requirements of processes and business functions. They may be supported by in-discipline specialists, such as high voltage specialists, building controls specialists and lighting specialists.

In the same way that mechanical and electrical engineers are lumped together as if they were the same, civil and structural engineers are often lumped together: only on very small projects might the personnel be the same.

■ Civil engineers are concerned with the ground outside. This revolves around the terrain levels and constitution and the drainage and roadways that run through the areas. They may be supported by in-discipline specialists as follows:
  ○ Land surveyors collect data to map the shape of land by means of topographic mapping. This is particularly important to building services engineers with respect to the cover of earth above utility services, especially if there is subsequent cut and fill, which may affect cover levels of existing utility services – as illustrated in Figure 4.3.
  ○ Geotechnical engineers are concerned with the physical and chemical properties of ground materials, including soil, rock and groundwater. As most building projects are supported by the ground, geotechnical engineers are involved with the design of foundations for structures, retaining walls, subgrades for roadways, embankments for water storage and flood control and containment systems for hazardous materials. In the worst case, geotechnical engineers also deal with various geological hazards impacting on society, such as landslides, soil erosion and earthquakes.
  ○ Traffic engineers design and manage roads to achieve the safe and efficient movement of people and goods. This includes the classification or road type, according to its use and speed, which determines the minimum lighting design criteria. Also,

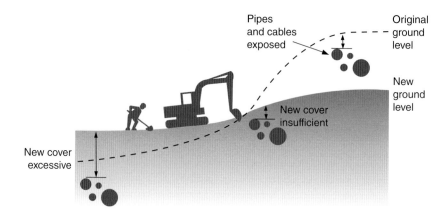

Figure 4.3 Effect of changing cover levels on utility services.

the location and operation of traffic signs, traffic lights and any other forms of control which will require electrical power supplies.

■ Structural engineers are responsible for ensuring that the design of buildings foundations, walls and roof are able to withstand loads and stresses that they will encounter from the weight of materials, people, wind and equipment. Building services equipment often relies upon the structural engineering solutions for their support and fixings or, as an integral part of their design. On other occasions it may be necessary to modify the structural engineering solutions to accommodate the requirements of the building services systems. Either way there may be a significant impact on the design of a building's structure and fabric or to the associated external works. Builders' work in connection (BWIC) is the work that needs to be done to order to allow for the building services installation to be completed, for example:
  ○ weight of plant
  ○ weight of water tanks full
  ○ details of pipe ducts in reinforced concrete floor
  ○ details of large holes in reinforced concrete floors and walls
  ○ details of plant supported on roof, such as roof fans.

Although BWIC is usually done by the building trades it needs to be well thought out at the earliest stages of design by building services engineers.

Quantity surveyors

Quantity surveyors are responsible for managing the costs on behalf of the client. This includes the construction costs and other payments associated with administering the contract, such as professional fees, land costs and taxes.

During the pre-contract stage, quantity surveyors assist in providing advice on procurement routes, preparing the tender documentation, receiving and analysing tenders and preparing the tender reports for the client and recommendations for approval. Where the client has a procurement department, quantity surveyors liaise with procurement throughout the procurement process.

During the contract period the quantity surveyors monitor the project spend, providing regular reports to clients, and they receive claims for payment for review. Quantity surveyors assist with negotiations associated with variations during the projects, which have a financial impact. Following completion of the construction works, the quantity surveyors will be involved with agreeing final accounts. It is normal for this to take up to 12 months after completion, but it could take longer if the project is complex and there have been many variations during the project.

Building services engineers need to provide the necessary information for costing. Under traditional methods of budget estimating and pre-contract cost planning, cost estimates for building services engineering generally were based solely on the gross floor area of a building. It may be inappropriate to use gross floor area as the sole descriptor for determining building services engineering out-turn costs because the building services engineering services costs do not have precise linear relationships with building form and building function:

■ with respect to costing variations, the costs are not simply associated with the costs of the materials and installation. They should include the necessary coordination of work between all the relevant members of the design team – and possibly site team if the works have started on site
■ owing to the shorter life cycle of building services engineering services compared with 'bricks and sticks', more life cycle costing considerations are needed. This is particularly significant in PFI type projects, where it is a significant consideration in the risk allocation.

Specialists

No one person can reasonably have complete knowledge, which is why specialist consultants play an important role in supporting the design team.

Cross-discipline specialists may either provide knowledge in a particular end-user area or provides a strategy level input across disciplines.

End-user area specialists provide details in areas which are not necessarily covered in standard guides and reference material and which may not necessarily be known to the design team members. They will help the design team understand how their designs will be used in practice. Examples are specialists who are related to:

■ end-user activities – equine, spa, theatre consultant – who will have an understanding of the peculiarities of that particular end-user and how these translate into requirements for the building services design

- operational activities, such as a catering consultant, who will understand the whole process from menu design, equipment and so on
- a particular aspect of the building design, suach as a specialist sports surfaces consultant.

Strategy level specialists in a multidisciplinary subject will determine the overall design strategy or philosophy for their specialism. This will inform the architects, engineers and cost consultants have to decide how to implement between them. This will inevitably involve trade-offs:

- Acoustics engineers are responsible for ensuring that the environment is free from nuisance sounds (ranging from annoying, which makes it difficult to function, to privacy concerns) and providing the appropriate level of speech intelligibility. Building services engineers are concerned with controlling noise from the building services equipment. Heating, ventilation and air-conditioning systems, hot and cold water services, process services, electrical generation systems, lift and escalator systems are all potentially the source of undesirable noise and/or vibration. This can either be addressed in the specification of the particular equipment (including the provision of acoustic enclosures) and/or modifications to the spaces to absorb, block, mask or redirect unwanted sound.

It is well known that building services create noise and vibration. Effective noise control of building services is not simply a question of locating attenuation devices at various points in a system. In an HVAC system, for example, the acoustic consultant must become completely familiar with the overall design. Only on this basis can comprehensive recommendations be made regarding duct velocities, damper locations, terminal device selections and potential duct borne crosstalk weaknesses as well as the basic requirement for plant room wall/slab thicknesses or even isolation of slabs in lightweight building structures.

- Fire and life safety engineers are responsible for developing strategies aimed at preventing, controlling and mitigating the effects of fires and smoke in line with statutory requirements for life safety and property protection goals. This involves identifying risks and design safeguards. The fire strategy will consider the building design and layout, human behaviour during fire events, emergency escape facilities and the behaviour of fire and smoke. The strategy will inform building services engineers on how to design the active fire protection system (fire suppression, fire detection and alarm system), smoke and heat control and management and emergency escape lighting. Whereas architects are concerned with the provision of passive fire protection (fire and smoke barriers, space separation).

- Security engineers are responsible for developing strategies for determining what needs protecting (people, vehicles, contents, building systems, building fabric), assessing the potential risks and defining mitigation methods. The people may include operational staff, maintenance staff and visitors to a building. This will need to consider structural engineering issues (buildings built with security in mind), technical (capitalising on electrical security technology), organisational (using professional staff, maybe animals, together with appropriate training) and procedural matters.
- Environment consultants and/or sustainability advisors are concerned with the environmental performance of projects. This includes the development, implementation and monitoring of environmental strategies, audits and assessments, policies and programmes that promote sustainable development, and also being pragmatic as to how these requirements integrate with other drivers such as cost (capital and operational) and programme. This includes establishing where improvements can be made and ensuring compliance with environmental legislation.
- Vertical transportation specialists provide advice and solutions to lift and escalator problems, site condition surveys, feasibility studies, maintenance contracts and lift refurbishment programmes for all types of passenger lifts, goods lifts and escalators. This may include a lift traffic analysis to determine the number, capacity and speed criteria for lifts in any building.
- Voluntary code assessors are responsible for ensuring that the process of accreditation of a scheme is properly managed. They provide expert advice on the process and information required and facilitate the team's efforts to successfully schedule activities, set priorities and negotiate the trade-offs required to achieve a target. They assess submitted evidence to certifying bodies for compliance.
- Disability access consultants identify how the access issues associated with a project are dealt with in the design, and identify the constraints that have had a bearing on the design.
- Wayfinding and signage consultants are concerned with the implementation of a good wayfinding system, which helps people to navigate a location by greeting, explaining and directing, using predetermined methods, including signs, maps, landmarks and written/verbal information.
- Waste management consultants are concerned with organisation and management of waste disposal, collection and recycling facilities. They may also be responsible for waste treatment and cleaning operations.
- Building physicists are concerned with understanding how the building, the occupants and the internal and external environments interact in terms of energy efficiency and building sustainability. They use the principles of physics applied to the sciences of architecture, engineering and human biology and physiology.

- Lighting consultants are concerned with designing the whole lighting system, including daylighting, to integrate with the architecture as well as the technical aspects. This may also include the design of bespoke luminaires.

Beware – there's no such thing as a free meal

Sometimes suppliers and manufacturers will offer technical consulting services whereby they will design systems for a project often at little or no apparent cost. These designs, of course, always incorporate only the products they sell. Under these circumstances, the supplier or manufacturer has a clear conflict of interest. Projects are better served by engaging truly independent specialists to design and specify a system to give you the best possible results, the greatest flexibility, and the best value for money.

4.4   The construction team

Construction teams provide the construction expertise, labour, materials and plant resources to deliver projects to the client's requirements – brief, budget and to programme. Construction teams comprise main contractors, subcontractors and suppliers.

Main contractors

Main contractors are responsible for day-to-day overseeing of the physical site works and management of suppliers and trade necessary to provide the final solution for the utility services to the premises. Depending on the particular contract, this may also include a level of design responsibility. In addition, the construction team will manage communication between all the involved parties throughout the course of project – applying for building permits and inspections, securing the site, providing temporary utilities, providing the workers under their control with relevant information and instruction, providing site surveying and engineering, disposal or recycling of construction waste, monitoring schedules and cash flows and maintaining accurate records – and the supporting functions which feed them: risk management, value management, information management, planning services, commercial management, quality management and staff development and training.

The key players that building services engineers will interface with are:

- project managers
- construction managers

- design managers, who are responsible for ensuring that there is a set of information that a building can be built from. This may involve acting as an intermediary between the design team and the construction team
- building services engineering managers
- health and safety personnel, who are responsible for health and safety on site and ensuring that all persons on site are aware of their responsibilities
- land surveyors, who are responsible for measuring and recording land
- project planners, who are responsible for coordinating the details of a project, creating a plan for executing the project and ensuring that all the individuals involved in the project are communicating and getting their assigned tasks completed according to schedule
- quantity surveyor/commercial managers, who administer the contracts
- document controllers, who are responsible for handling and organising information.

Subcontractors

Main contractors may subcontract part of the work, to relieve themselves of part of the building works, or to expedite it at a lower cost or at a greater or more skilled level. It also provides a means of transferring risk. Main contractors will be responsible for the subcontractors performance, and will coordinate their work accordingly. Subcontractors may then subcontract elements of their work further.

Subcontractors deliver their selected work packages, in coordination with main contractors and other subcontractors, particularly in the case of building services subcontractors whose work generally requires significant coordination with other subcontractors.

Suppliers

Suppliers provide goods and/or services to main contractors or subcontractors. These may either be a direct purchase from their factory or via distributors and/or wholesalers.

4.5   Utility service providers

There are four aspects associated with providing utility services to buildings that need to be considered: the network that the building's utility services will tie into, the physical equipment and apparatus required for the tie-in to the network, the provision of the utility service at the point of supply and its metering. Each aspect may be provided by the same or different organisations. These arrangements are subject to change as organisations merge, or new organisations enter the market.

Privatisation of utility service provision in the UK

Historically utility services were provided by state-run monopolies. During the 1980s and 1990s, due to political drivers these were progressively deregulated and broken up, with the aim of providing better services through competition and relieving the state of having to provide these services directly. This has led to a situation where the progress of the design of the utility services is frequently hampered by unreliable record information and dealing with multiple authorities, for example, gas transporters, shippers and suppliers, electricity suppliers, distributors and meter operators, water suppliers or undertakers and telecoms network and service providers.

Network operators are responsible for owning and operating the distribution and transmission infrastructure that delivers the utility service.

The connection installer provides the physical interface and any reinforcement to their existing network. The physical issues include planning for and installing materials such as cables, pipes and associated apparatus for both public and private utility services, coordinated with each other and with other physical encumbrances and legal constraints usually in a live environment as illustrated in Figure 4.4.

Utility service suppliers make available and sell the utility service to the consumer in accordance with defined tolerance limits. Meter suppliers, supply, install, maintain and read the meter and bill and collect the revenues.

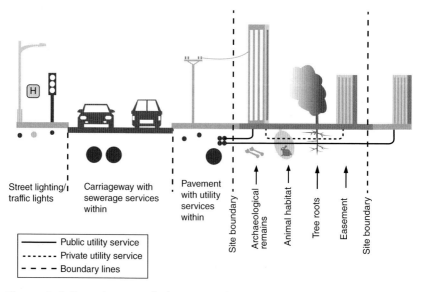

Figure 4.4 Physical aspects of utility services design.

4.6   Non-contractual interfaces

There are potential interfaces with parties that have no contractual link to the project but have interests that building services engineers will need to recognise.

- The general public, neighbours to a project and the media may voice views (positive or negative) on matters such as external lighting and appearance and noise from building services equipment.
- Awards bodies may determine building merit recognition for some aspect of the building services design, such as sustainability, innovation or implementation of systems. These may either be planned for, as part of the design, or simply awarded retrospectively.
- Pressure groups may be striving for or promoting changes, to influence public opinion and/or policy in relation to issues such as environmental concerns.
- Industry trade associations are founded and funded by businesses that operate in a specific industry and therefore have no statutory authority, but they can provide valuable information.
- Watchdogs who set standards and regulate behaviour.
- Industry regulators administer government legislation, particularly to ensure that the public receives adequate quality of service, that there is fair competition and to impose sanctions for non-compliances.

In addition, there are professional institutions. These are non-profit organisations seeking to further a particular profession, the interests of individuals engaged in that profession and the public interest. They may have a number of functions such as:

- setting standards and assessing prospective members for entry
- providing support for continuing professional development (CPD) through learning opportunities and tools for recording and planning
- publishing professional journals or magazines to inform members of the latest developments
- organising networks and conferences to allow for dissemination of contemporary knowledge
- issuing codes of conduct to guide professional and ethical behaviour
- dealing with complaints against professionals and implementing disciplinary procedures
- producing authoritative guidance material
- developing and promoting certification and accreditation schemes.

IEE Wiring Regulations in England and Wales

The most important document produced by the IET is the IEE Wiring Regulations which set the standard for electrical installation in the UK, and in many other countries. Colloquially referred to as 'the Regs' they became BS 7671, so that the legal enforcement of their requirements is easier, both in connection with the Electricity at Work Regulations and from an international point of view. They require that installation owners and their designers consider health and safety requirements during design and construction and throughout the life of an installation, including maintenance, repair and demolition. They are intended to provide safety to people or to livestock from fire, shock and burns in any installation which complies with their requirements, but they do not cover public utility supplies.

They are not intended to take the place of a detailed specification, but may form part of a contract for an electrical installation. The Regulations themselves contain the legal requirements for electrical installations, while the Guidance Notes indicate good practice.

In Scotland, the IEE Regulations are cited in the Building Regulations, so they must be followed. While failure to comply with the Regulations has not generally been a criminal offence, those who complete such installations may be liable to prosecution in the event of an accident caused by faulty wiring.

BS 7671 has the status of a British Standard code of practice, and is thus not a statutory instrument. However, the standard is a benchmark for the design of low voltage electrical installations in the UK. In Scotland, it is included in the Building Control requirement. The Memorandum of Guidance to the Electricity at Work Regulations, which makes reference to the British Standard, notes that installations that are designed in conformity with the British Standard are likely to achieve compliance with the Electricty at Work Regulations 1989. The HSE's electrical inspectors generally use BS 7671 as their main guide when checking electrical installations, and BS 7671 is commonly used in civil litigation concerning safety of electrical installations.

Low Carbon Energy Assessors and Low Carbon Consultants

LCCs are members of the CIBSE Low Carbon Consultants Register and are professionals competent to minimise energy use and carbon emissions from buildings in design, operation and simulation. They are able to go beyond the current minimum legal requirements in improving the energy performance of both new and existing buildings.

Clients can be assured that buildings designed and operated by LCCs will meet the requirements of Part L (Conservation of Fuel and Power), and Building Control Officers can be assured that compliance is being signed off by suitably qualified professionals. Members of the LCC Register have undertaken assessments to demonstrate their competence.

Summary

For any project building, services will have to interface with a plethora of other stakeholders. Relationships between organisations and individuals will often be formed for the duration of a project and subsequently disbanded. They may or may not be re-formed for future projects. Some relationships are direct contractual lines, others will be via intermediaries, and some will have no contractual lines.

Client teams need competencies in health and safety, legal, financial/commercial and operational areas. It is normal to divide these so that the appropriate management structure can be implemented and conflicts of interest avoided. For some projects, clients may not know the identity of the end-user – for example, when the space is built to rent – thus, letting agents will also be required.

Reference

BSRIA (2013) *Legislation and Compliance*. Available: https://infonet.bsria.co.uk/legislation-compliance/ [accessed September 2013].

5 Professional ethics

Building service engineers require specific competencies and skills to ensure that their work complies with minimum statutory requirements; however, their work has a direct and vital impact on the quality of life in the built and natural environments. In this context, professional ethics can be considered to be the responsibilities and obligations to the general public, clients, employees and the professions, over and above any minimum statutory requirements. There are personal ethics and organisational ethics.

Values and morals

Values, such as honesty, integrity, responsibility, respect and fairness, are some of the principles used to define what is right, good and just. They provide guidance to set the standards in determining the right versus the wrong, good versus the bad and just versus unjust.

Morals are values attributed to a system of beliefs. These values get their authority from something outside individuals – a higher being or higher authority, such as societal beliefs associated with sustainability. However, the ultimate moral driver is the value of human life imposing the highest moral considerations from those who might otherwise put it at risk.

Personal ethics are driven by a matter of personal pride, professional integrity and conscience. Also, ethical codes of conducts are published by professional institutions. These will state guidelines for the conduct of members in fulfilling their obligations to other professionals, clients

Building Services Design Management, First Edition. Jackie Portman.
© 2014 John Wiley & Sons, Ltd. Published 2014 by John Wiley & Sons, Ltd.

and the general public, both in the short term and in shaping the built environment for the future.

Organisational ethics are driven by pressure to consider the 'triple bottom line', which adds social and environmental concerns to commercial demands. Balances needs to be addressed between culture and commerce, employment and conservation, health and cost, private and public interest and heritage and utility. Potential barriers are overriding commercial priorities, lack of any consequences of not complying, supplier pressures, ineffective training and regulatory compliance. Some aspects of organisational ethics – such as sustainability and contribution to educational and social programmes – may be documented in a 'corporate social responsibility' statement and/or in an ethical code of conduct.

The balance between personal and organisational ethics may cause tension. To coalesce personal and the organisational values organisations need to address the professional obligations of their employees. The prioritisation of obligations for individuals can be a difficult task because if personal value dominates over organisational value, business would no longer be viable and if the organisational value dominates, then the likelihood of unethical and illegal conduct increases.

Examples of possible organisational ethical decisions: Areas of work

An organisation may decide not to solicit work in certain areas, such as those associated with the design of animal testing facilities, associations with a particular industry (defence or nuclear) or product (cigarettes, alcohol or weapons) or may not solicit work in certain countries with oppressive or corrupt regimes or poor human rights records.

Ethics is about actions and decisions: by acting in ways that are consistent with beliefs, we demonstrate ethical actions. When actions are not matching to values – our sense of right, good and just – we will view that as acting unethically. Ethical breaches are manifested in the form of unfair conduct, negligence, conflict of interest, collusive tendering, fraud, breaking confidentiality and propriety breach and bribery. The main types of unethical behaviour that building services engineers could be susceptible to are:

- concealing known errors, withholding information, distortion of facts
- exaggerating or misleading experience and academic achievements in CVs and applications for commissions
- charging clients for work not done, costs not incurred or overstated
- false promises of advancement of works

- 'by-standing' – failing to intervene or report wrongdoing within an area of responsibility (this does not give licence to interfere anywhere and everywhere, which is itself unethical for various reasons)
- conflict of interests (having a foot in two or more competing camps)
- breaking confidentiality
- neglect of duty of care.

Moral dilemmas

Dilemma: Is taking meals and entertainment with potential suppliers unethical, as it may influence decisions on specifications?

Discussion: There is a dual standard here. Meals and entertainment are an important part of forming a 'partnering' relationship and this is a good, ethical way of doing business; others may see this as the same thing as 'inducement' … there is merit to both arguments and, possibly, policies that avoid extremes are likely the best.

Dilemma: When considering energy consumption tenants pay the running costs of the building but the general public (present and future generations) bears the price of depleted fuel stocks, greater pollution, CO_2 emissions etc. Where should building services engineers align their efforts?

Discussion: The person or organisation who pays the fees are clients who may want to maximise future income and minimise capital costs, for example, by reducing the specification of the thermal insulation and of the controls for environmental services. At some point a happy balance needs to be agreed upon.

Ethics also acts as a risk management tool. If ethical standards at any level of the organisation weaken – particularly within top management, which sets the tone for the rest of the organisation – it can trigger a domino effect of unprofessional practice that will ultimately harm the entire organisation.

Summary

Statutory requirements and codes of practice can provide a framework to base ethical standards on: they identify the moral norms and render them more explicit. However, often it is how the rules are interpreted and how an entity, acting freely but within professional boundaries, decides what is 'right'. Lapses can lead directly to losing the privilege to operate and the associated financial consequences. In the long run, the cost of unethical working outweighs the benefits of any short-term gains realised.

Part Two Technical issues associated with building services design

6 Design criteria

Design criteria set out the parameters that the building services engineering systems should be designed to, and against which the installation can be evaluated. Building services engineers are concerned with both the external and internal design criteria – with the building fabric acting as a buffer and climatic modifier between the outdoor and the indoor environments.

Building Services Design Management, First Edition. Jackie Portman.
© 2014 John Wiley & Sons, Ltd. Published 2014 by John Wiley & Sons, Ltd.

6.1   External design criteria

External design criteria define the environment in which a building is built. Although the external design criteria are beyond their control, building services engineers need to specify a range of benchmarks, as a basis for the design, within which it will be possible for the defined indoor design criteria to be achieved. Not using the correct outdoor design conditions can cause errors that will propagate throughout the building services design process, resulting in an uncomfortable indoor environment and reduced energy efficiency.

There are also cost considerations. Ideally, the design of the building should ensure that the desired interior design criteria are achieved, irrespective of any conditions outside the building. However, this is not realistic and the areas of the graph in Figure 6.1 illustrated this:

(1)   The cost of the building services design and installation would be almost infinite in order to guarantee being able to achieve precisely defined indoor conditions irrespective of *any* external conditions. This would need to survive colder and hotter temperatures, more rainfall and higher winds than currently recorded: in reality this should be considered an impossibility.

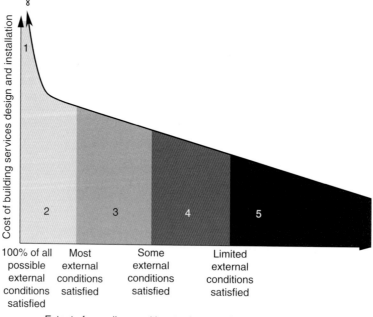

Figure 6.1 Relationship between cost and satisfying external design criteria.

(2)   This segment represents the range of external design criteria within which it is possible to maintain the desired indoor conditions – at a particular cost.

(3)   At a lower cost, it is not possible to maintain the same level tightness of internal design conditions, so the interior is tolerable for the occupants, by means of certain adjustments such as:

- making personal adjustments such as clothing, eating/drinking hot/cold food or beverages
- making local environmental adjustments such as opening/closing windows
- making operational adjustments such as, rescheduling activities, relocating away from or closer to windows
- managing expectationssuch as understanding and appreciating that on very hot days it will be warmer inside.

(4)   If the cost of the building services engineering installation is even less, then the building may only be tolerable – with respect to the internal environment – for some of the time or for certain aspects only: it may not be possible to make adjustments to ensure that all the interior design criteria are acceptable at all the times required.

(5)   Obviously, if insufficient cost is allowed – at the extreme, nothing – it will be impossible to satisfy any comfortable interior design criteria.

This particularly applies to indoor temperature and humidity, but the same principle applies to lighting, and other indoor design criteria.

As well as maintaining interior design conditions for people, consideration is also required for maintaining the interior design criteria for processes. These may be less or more onerous than for people. The most onerous will prevail.

An important part of the building services engineers' role is to manage the client's expectations with respect to the costs and different risks of exceedance from the desired interior design conditions. This involves trade-offs between the end-users' expectations and tolerances and the costs.

Weather and climate

The difference between climate and weather is subtle but important. Climate is what is expected, based on how the weather changes over a long period of time, typically over 30 years or maybe longer. Weather is how the atmosphere is behaving at a moment in time. Building services engineers will base design criteria on climatic predictions, but the finished building will need to deal with the actual weather.

The key external design criteria which building services engineers have to consider are meteorological and pollution factors, from both the built and natural environment, in the particular locality.

■   Meteorological factors comprise of temperature and humidity, solar radiation (global, direct and diffuse), rainfall, wind (speed and direction) and the risk of lightning strike. All of these may be modified by

the features of the local environs (topography, ground roughness, nearby obstacles) as well as seasons and weather extremes, for example, hurricanes, droughts or rainy periods.
- Pollution factors comprise air quality, noise and vibration, light and vermin which cause contamination: these may either occur naturally or be human induced.

The milieu of a particular location tends to influence the shapes and forms of the local buildings and dictate the types of environmental control required. Building services engineers are particularly concerned with those climatic variables which affect the potential to achieve the desired indoor thermal comfort and the building energy efficiency. This relates to the heat transfer through the building fabric and via ventilation, and the consequential effect on the sizing of the building services engineering plant. It is more than a coincidence that building designers in different parts of the world appear to have come up independently with similar building design solutions in their endeavours to overcome the local prevailing unfavourable external weather conditions. There is often a distinct correspondence between special architectural features and different climatic zones, as these examples show:

- In temperate climates heating requirements are significant, and the main consideration of thermal design is to minimise winter heat loss through appropriate thermal insulation to the external walls and the use of double and sometimes triple glazing for windows.
- In tropical climates heat gains and hence cooling requirements will be the main design consideration. Space heating is either not required or is insignificant. Since conduction heat gain through the external walls usually accounts for a small percentage compared with solar heat – which is the major component of building envelope heat gain – thermal insulation to the external walls is often not cost-effective. Shading devices and/or glazing with small shading coefficients such as tinted and reflective glasses is commonly used.
- In polar climates the ambient temperature reaches extreme values and this has a large impact on the heat loss through the building envelope. The solar pattern is completely different: there is limited availability in winter, yet in summer the sun is above the horizon for 24 hours. Strong winds and fierce storms have a major effect on the infiltration of buildings and they heavily influence the infiltration heat loss through the building envelope. The wind patterns have huge influence on the local microclimate around the building and create snowdrifts and the problems with thawing, icing and possible condensation in the building envelope. The humidity in the interior is driven out through the building envelope in the winter due to the pressure difference, strong winds and low water ratio in the outdoor air. Designs are characterised by increased insulation in a super airtight building shell, super

efficient windows to produce the net positive solar gain, and a
ventilation system with very efficient heat recovery.

Climate change

There is a plethora of 'advice' anticipating and predicting changes,
much of which will affect the design of buildings. These vary in their
complexity and can only give probabilistic projections. The headline
climate impacts currently predict hotter drier summers, warmer wetter
winters, rising sea levels and an increase in the number of extreme
weather events. Some of these are being translated into codes, regulations and guidance used by building services engineers.

Meteorological design criteria

Building services engineers need a representation of the weather in a
particular location and usually rely upon published weather data based
on long-term records for particular areas. It is important to understand
the basis of the data used: the accuracy and reliability of the data will
depend on the length of time and the quality of the measurements. The
longer the period of records and the more recent the weather data, the
better and more representative the results will be. Since shorter periods
may exhibit variations from the long-term average, only considering a
very early period of data may not reflect the present weather conditions.
For building design, weather data based of not less than 25 years is considered to be conservatively stable.

For example, CIBSE can provide weather data files for 14 locations
around the UK, as illustrated in Figure 6.2. The area represented by each
weather data file varies considerably. Due to the limited availability of
data, some areas of the UK are better covered than others, and for a
country with such large variations in temperature over relatively small
distances, this can present a problem.

Building services engineers, in consultation with the other design
team members, need to agree and record the design assumptions pertaining to the external design criteria. This data is representative of the
average climate over many areas but it does not necessarily give an indication of the range of possible conditions. While this is not an issue for
determining average energy use or typical carbon emissions it may be
an issue when considering peak heating loads, human comfort or overheating levels.

Temperature and humidity

The outdoor design criteria of temperature and humidity are used
when calculating the building's thermal (heating and cooling) loads.
These usually comprise the dry-bulb and wet-bulb temperatures and
relative humidity. If incorrect assumptions are made, this affects the
sizing, optimal equipment selection, distribution and installation of

Glasgow

Edinburgh

Belfast

Newcastle

Leeds

Manchester

Nottingham

Birmingham

Norwich

London

Cardiff

Southampton

Swindon

Plymouth

Figure 6.2 Weather data stations.

heating, ventilating, air-conditioning and dehumidification equipment, as well as other energy-related processes. Table 6.1 gives a typical sample specification.

The dry-bulb temperature refers to the ambient air temperature and is usually referred to as air temperature; this is the air property that is most commonly used. Dry-bulb temperature can be measured using a normal thermometer freely exposed to the air but shielded from direct radiation and moisture – hence the term dry-bulb.

The relative humidity is the amount of moisture in the air as a percentage of the most moisture that could be held in the air at a given

Table 6.1 Typical sample specification for outdoor temperature and humidity.

Season		Design criteria
Summer	For sizing cooling installation	31 °C, dry-bulb, 21 °C wet-bulb
	For naturally ventilated area	Overheating criteria as set out in CIBSE (2005), AM10 in Nondomestic buildings
Winter	For sizing heating	−5 °C/100% RH
	For sizing protective installations, for example trace-heating	−15 °C/100% RH
	Air frost projection coils	−10 °C/100% RH

temperature. If the air has half the amount of moisture it could have then the relative humidity is 50%. When it is raining or snowing and the maximum amount of moisture has evaporated into the air then the relative humidity will be 100%. The humidity of the air is important as it will affect the efficiency of cooling and heating systems having to deal with modifying the outside air. Also, any water vapour which enters the building may condense to a liquid which can cause problems.

The measured wet-bulb temperature is a function of relative humidity and ambient air temperature. The difference between the wet-bulb and dry-bulb temperatures shows how much water vapour the atmosphere currently holds. A lower wet-bulb temperature means that the air is drier and could hold more water vapour (low relative humidity); a higher wet-bulb temperature means that the air cannot hold much more vapour (high relative humidity).

Wet-bulb temperature is particularly relevant when designing evaporative cooling systems such as in cooling towers, since the wet-bulb temperature, by its very nature, represents the limit of performance for evaporative cooling as the wet-bulb temperature is achieved by evaporation of water, and is the lowest possible temperature an evaporative cooler could produce. Actual performance of evaporative cooling depends on the effectiveness of the particular unit, expressed as a percentage of the possible temperature drop, which is the difference between the dry-bulb and wet-bulb readings.

Solar radiation

Radiation from the sun affects the building's orientation, window placement and glazing specifications and opportunities for daylighting. Knowledge of solar radiation is required to determine the cooling loads and for solar energy applications, such as for passive solar design (where the building fabric collects, stores and distributes solar energy in the form of heat in the winter and rejects solar heat in the summer), solar water heating and photovoltaic electricity generation. Also, high intensities of solar radiation will degrade certain materials, so cognisance may be required of any externally located building services equipment.

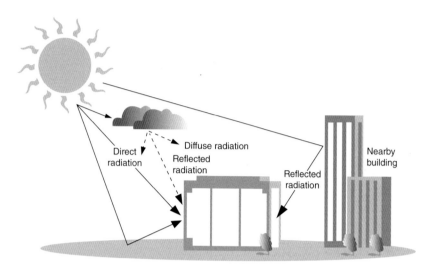

Figure 6.3 Elements of solar radiation.

As illustrated in Figure 6.3 the sun's rays will arrive at a building by (a) direct radiation, which is an unimpeded direct line from the sun, (b) diffuse radiation, which is light scattered by atmospheric constituents, such as clouds and dust, and (c) light reflected from the ground and other surfaces. The sum of the direct, diffuse and reflected radiation is called total or global solar radiation.

With respect to the direct radiation, sun path diagrams represent the course of the sun across the sky at different times during the day and throughout the year. Nowadays 3D computer graphics are used to determine location-specific direct radiation as it varies on a seasonal basis. This can also model the effect of shadows cast by buildings, trees and landforms on and around the site. The output will typically be in an animated, rotatable, 3D colour graphic model of a proposed building design. The design criteria should specify which software and database is being used and should have been validated by a suitable professional body.

Rainfall

The amount, direction and intensity of rainfall will inform aspects of a building design such as the roof form with respect to rainwater collection, storm water drainage and potential for rainwater harvesting and sustainable urban drainage systems (SUDS). In the UK, the Met. Office maintain a database of information with rainfall measured by rain gauges and sensors. Table 6.2 contains typical design criteria.

Ice and snow

Ice can cause problems in the cooling of equipment or freezing and thawing, which will result in cracks occurring, breaking cases and pipes. Powdered snow can be blown through ventilation ducts and then melt

Table 6.2 Typical sample specification for rainfall.

External design criteria for rainfall
Based on the information from the UK Met. Office, the following will
   used as the basis of design:
- average annual rainfall: 80 mm
- maximum recorded annual rainfall/month: 302 mm (January)
- minimum recorded annual rainfall: <1 mm

Table 6.3 Typical sample specification for wind.

External design criteria for wind
Prevailing wind direction is from north-west to north.

Height (AGL)	Average speed
10 m	4.6 m/s
25 m	5.5 m/s
45 m	6.0 m/s

in equipment compartments and cubicles, which can cause damp problems in critical systems if not prevented in the original construction.

Wind

The average prevailing wind speed and wind direction data is particularly relevant when natural ventilation, air exchange and infiltration
wind pressures are an issue. Wind speed and frequency may also be a
factor in selecting wind as a power generation source, and extreme
annual design wind speeds are pertinent in designing smoke management systems. In the UK, the Met. Office maintains a database of wind
speeds as measured using weather vanes and anemometers. Table 6.3
contains typical design criteria.

Weathering

Certain organic materials; for example, plastics, rubbers and paints will weather
due to the combined effect of solar radiation, temperature and humidity changes
and contaminants. This are manifested as:

- rapid deterioration and breakdown of paints
- cracking and disintegration of cable sheathing
- fading of pigments
- bleached out of colours in paints, textiles and paper, a major consideration
  when trying to read the colour coding of components such as cable
  insulation.

Air pressure

Air pressure, frequently referred to as atmospheric pressure, is 'the force exerted on a surface of unit area caused by the earth's gravitational attraction on the air vertically above that area'.

Air pressure varies with altitude

It is not widely appreciated that the location of building services equipment, especially withrespect to its altitude above sea level, can affect the working of that equipment.

But it is not just the height above sea level that has an effect. Even air pressure variations at ground level have to be considered.

At altitudes above sea level, low air pressure can cause:

- leakage of gases or fluids from gasket sealed containers
- ruptures of pressurised containers
- change of physical or chemical properties, for example for diesel fuel
- erratic breakdown or malfunction of equipment from arcing or corona
- decreased efficiency of heat dissipation by convection and conduction in air, which will affect equipment cooling; for example, an air pressure decrease of 30% has been found to cause an increase of 12% in temperature
- acceleration of effects due essentially to temperature, for example volatilisation of plasticisers, evaporation of lubricants.

Risk of lightning

A lightning strike can cause damage due to the intensely high voltage currents striking the structure and causing either primary or secondary physical damage or impairing sensitive electronic components owing to surges in the electrical distribution system.

The risk of a lightning strike to a building is a function of the size, type of structure, height, number of lightning strikes per year per m^2 for the region, and the proximity of other buildings. The impact of a lightning strike to a building is assessed with respect to potential loss of human life, loss of service to the public, loss of cultural heritage and loss of economic value. Together these determine the extent of the lightning protection measures needed to protect both the outside of the building and the sensitive equipment within. Table 6.4 contains typical design criteria.

Table 6.4 Typical sample specification for lightning protection risk.

External design criteria for lightning risk

A lightning risk assessment, undertaken in accordance with to BS EN 62305-2 has determined that a lightning protection system shall be provided to Class III.

Other potential external design criteria

Other specific external design criteria which affect the design of the building services systems may be included if necessary, such as:

- risk of flooding
- potential for seismic activity and earthquakes
- likelihood of sand and dust storms.

If the building is to be located in an ultra-challenging environment – for example, underwater, deep underground, on the moon or in space, then the same principles of establishing external design criteria apply.

Microclimates

The data available to building services engineers may not truly reflect local conditions: many metrological data collection sites are some distant from the location of the building, and there may be local microclimate issues. Microclimate issues may refer to areas as small as a few square feet (for example, a garden bed) or as large as many square miles. Some examples are given below:

- Proximity to bodies of water – with reference to Figure 6.4 moving air will pick up moisture and will heat the water evaporating from the lake. When this reaches the land the warmer air is less dense, which will cause it to rise. As the temperature decreases, the water vapour will condense and precipitate and, due to gravitational pull, fall as rain. If the temperature is very low, snow may fall.

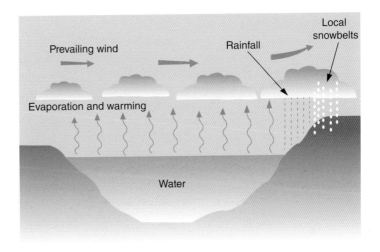

Figure 6.4 Effect of water on local weather.

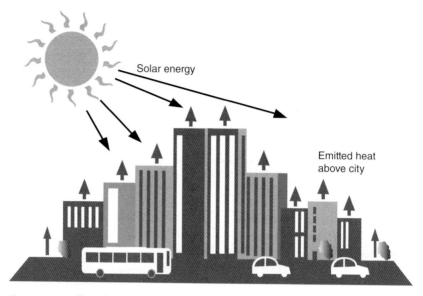

Figure 6.5 Effect of heavy urban areas.

Other effects are local increases in cloudiness and an increase the risk of flooding.

■ Coastal areas and off-shore islands – these have a high saline atmosphere, often accompanied by high salt content in the ground. Very often the ground water level is only just below the surface and may vary with the tide, even though the coast may be several miles away. There may also be coastal sand. A saline atmosphere and effects of sandstorms may attack exposed surfaces and damage the internal parts of motors, fans and control equipment if it is not suitably protected.

■ Heavy urban areas – these have a higher ambient temperature than nearby rural areas due to the absorption and subsequent re-radiation of mainly solar heat to the ambient air from buildings and roads and the human and industrial activity of urban areas, as illustrated in Figure 6.5. This necessitates an increase in the amount of energy used for cooling purposes, and also increases pollution.

This 'urban heat island effect' leads to increased rainfall, both in amounts and intensity, downwind of cities.

■ The slope or aspect of an area – south-facing slopes in the northern hemisphere (and north-facing slopes in the southern hemisphere) are exposed to more direct sunlight than opposite slopes and are therefore buildings are subject to greater heat gains.

■ Local wind patterns – buildings and other obtrusions above the ground can influence local wind patterns in several ways: some act

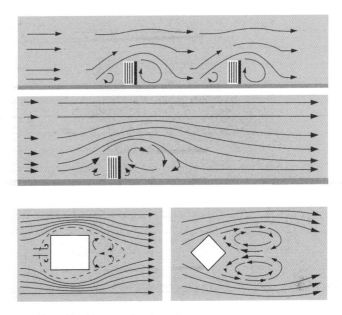

Figure 6.6 Effect of buildings on local wind patterns.

to increase wind speed, others to decrease it, and all acting to alter wind directions from the overriding weather pattern, as illustrated in Figure 6.6. Wind patterns are particularly important with respect to natural ventilation systems.

■ Local soil conditions – Different soils have different abilities to retain heat, which is relevant in the design of ground-source heating. The ability of soil to retain water will affect the design of storm water drainage systems and irrigation systems.

■ Local topography – An undulating landscape affects the temperature and humidity of air in the locale; for example, cold air holding water vapour will gravitate to local hollows or blockage points, leading to local fogging and ultimately greater condensation, as illustrated in Figure 6.7.

Pollution and contaminants

Building services engineers have very little control over the nature of existing pollution in the vicinity of the building. However, the design should ensure that the new building and the process of construction does not contribute to permanent additional pollution issues. Any interim additions to pollution during construction are managed to be within any statutory requirements.

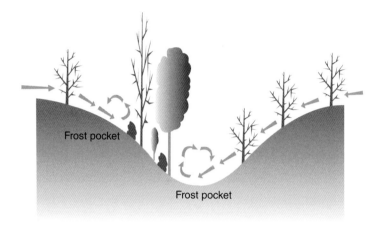

Figure 6.7 Local fogging.

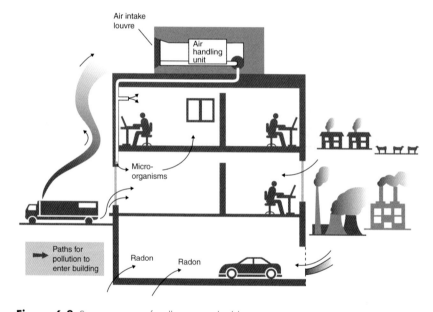

Figure 6.8 Some sources of pollution into buildings.

Air quality

Air pollution is the process of introducing contaminants (air borne chemicals or compounds: dust, sand, smoke and other particles) into a natural environment. These can be harmful to human health or well-being. Some are illustrated in Figure 6.8.

Sources of natural pollutants include:

■ sulphur – emitted by volcanoes and arising from biological processes
■ nitrogen – from biological processes in soil and lightning and bio-mass burning

- hydrocarbons – methane from fermentation of rice paddies, fermentation of the digestive tract of ruminants, such as cows, and also released by insects, coal mining and gas extraction
- radon, which occurs naturally in some soils.

Sources of man-made pollutants include:

- carbon dioxide and carbon monoxide produced during the burning of fossil fuels
- soot formation accompanied by carbon monoxide and generally due to inadequate or poor air supply
- hydrocarbons – most boilers and central heating units burning fossil fuels have very low emissions of gaseous hydrocarbons or oxygenated hydrocarbons such as aldehydes
- dust.

As pollutants move from one medium to another they may be deposited on building services equipment and equipment housing; they can cause extensive damage. They can have an effect on building services equipment in various ways, especially:

- ingress of dust into enclosures and encapsulations
- deterioration of electrical characteristics (e.g. faulty contact, change of contact resistance)
- seizure or disturbance in motion bearings, axles, shafts and other moving parts
- surface abrasion, erosion and corrosion
- reduction in thermal conductivity
- clogging of ventilating openings, bushes, pipes, filters and apertures that are necessary for operation.

The presence of dust and sand in combination with other environmental factors such as water vapour can also cause corrosion and promote mould growth.

Damp hot atmospheres cause corrosion in connection with chemically aggressive dust, and similar effects are caused by salt mist. Effects of ion-conducting and corrosive dusts – for example de-icing salts – need also to be considered.

The main requirements for air quality are in relation to compliance with national standards which are long-term benchmarks for ambient pollutant concentrations which represent negligible or zero risk to health.

Noise and vibration

A new building must pay due regard to the existing noise conditions in order that future occupants do not suffer from excessive interior noise levels and to ensure that the completed building does not impose additional external noise sources on the surrounding environments. Similarly, when

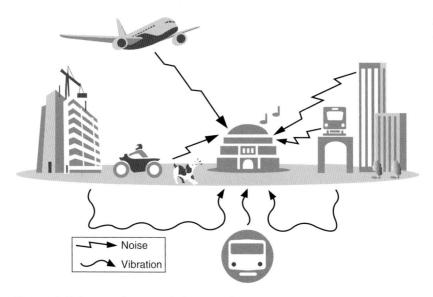

Figure 6.9 Sources of noise and vibration pollution.

a building is close to a source of vibration, such as a railway or under-ground train line, the existing vibration levels need to be considered. Figure 6.9 illustrates possible sources of noise and vibration pollution.

It is necessary to measure and record a baseline noise survey to quan-tify the existing noise levels present in an environment. The noise level may be measured at any moment, but it will vary widely with time, such as with the coming and going of vehicles and aircraft. Therefore, a single measurement says very little about ambient noise.

Complete noise histories can, however, be recorded graphically, but such charts are unwieldy and cannot be condensed to fit easily into reports, and they are difficult to interpret and describe verbally. The report usually calculates the equivalent continuous noise level which has the same energy as the original fluctuating noise for the same given period of time and percentile levels which are exceeded for a percentage of a stated time period. Percentile levels reveal maximum and minimum noise levels.

Light

A new building may include new external lighting or necessitate changes to street lighting and, as a result, there will be changes to the existing baseline background lighting which could cause nuisances to people or animals. So it is important to ascertain the existing baseline lighting conditions on the site and in the immediate surroundings prior to the design commencing.

Figure 6.10 illustrates potential sources of light pollution:

■   light spill: the unwanted spillage of light onto adjacent areas, which may affect sensitive receptors, particularly residential properties and ecological sites

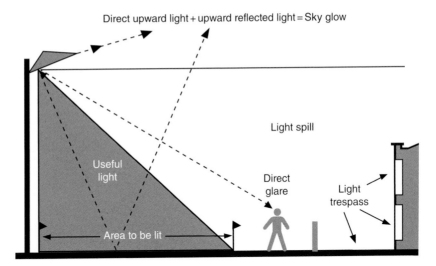

Figure 6.10 Sources of light pollution.

■ sky glow: the upward spill of light into the sky which can cause a glowing effect and is often seen above cities when viewed from a dark area
■ direct glare: the uncomfortable brightness of the light source against a dark background which results in dazzling the observer; this may cause nuisance to residents and a hazard to road users
■ light trespass (into windows): the spilling of light beyond the boundary of a property.

The background light readings should be recorded at key locations to benchmark the current nocturnal scene conditions. This consists of a night-time survey where readings are taken of the existing ambient lighting levels and observations are made of the existing lighting installations on the site and the surrounding area. Also sensitive receptors (people and animals) in close proximity to the site should be identified using professional judgement.

Fauna

With a few exceptions, fauna (rodents, insects, termites, birds etc.) may be present at all locations where building services equipment is present. Where they are a nuisance they are classed as vermin.

Vermin can cause a nuisance as a result of their faeces (being acidic and very damaging to building services equipment), eating through insulation on cables – possibly leading to short-circuiting electrical equipment, or in the case of birds, building nests and deposited feed stocks (often of flammable materials, leading to a fire hazard) and metabolic products such as excrement and enzymes which can become a fire hazard.

Larger vermin can also cause mechanical damage which can cause the physical breakdown of equipment and electrical failure caused by mechanical deterioration. Mechanical damage to wiring systems by larger animals such as cattle and horses can often be prevented by careful siting of equipment and physical barriers.

Any vermin in ductwork, rain gutters and drains can cause damage to the operation of the systems. The presence of such hazards could be identified by a visual survey. Weatherproof louvres may provide some protection against leaves and other fauna entering a building.

Flora

Flora (plants, trees, seeds, fruit, blossom, mould, bacteria and fungi etc.) may be present in open air locations where building services equipment may be located, while moulds and bacteria may be present both inside buildings and in open-air conditions. Deposits from flora may consist of detached parts of plants – leaves, blossom, seeds and fruits, and growth layers of cultures of moulds or bacteria. This can lead to deterioration of material, metallic corrosion, mechanical failure of moving parts and electrical failure due to increased conductivity of insulators, failure of insulation, increased contact resistance, electrolytic and ageing effects in the presence of humidity or chemical substances, moisture absorption and adsorption and decreased heat dissipation. These in turn can cause interruption of electrical circuits, malfunctioning of mechanical parts and clouding of optical surfaces (including glass). They may also cause blockages in drains, grilles and guttering or clog up access to services via access panels.

A particular case of flora contamination is the Japanese knotweed. This is a giant herbaceous perennial which can grow up to 10 cm per day in any type of soil. It forms dense clumps of up to 3 m in height. It thrives on disturbance and spreads by natural means and by human activity. Reproduction occurs both vegetatively (rhizomes) and through seeds, making this plant extremely hard to eradicate. Very small fragments of rhizome (the underground stems), even the size of a coin, can produce fresh new plants. Below ground each stand creates a rhizome network that can extend to 3 m in depth and 8 m in all directions. This makes it a serious threat to construction where it can have devastating consequences damaging foundations, drains and other underground services.

6.2   Interior design criteria

The interior design criteria should be specified to ensure that the indoor conditions satisfy the occupants and/or processes in the different internal spaces. These will be used as a basis for verifying the design during the commissioning stage.

Some indoor design criteria, particularly where driven by health and safety criteria, are covered by minimum statutory requirements. These may be enhanced by clients, as a preference or as part of a desire to comply with particular voluntary codes and practices and by the design team, by virtue of good engineering practice.

Building services engineers are principally concerned with thermal, visual and acoustics comfort, electromagnetic fields and static electricity. Occupants of buildings are becoming more conscious of and critical of the aspects of the indoor environments provided by the building services systems. What constitutes a comfortable environment is a deceptively simple question with profound implications for building services engineering designers. While the perception varies according to tangible parameters such as age, gender, individual's metabolic rates, state of health and clothing, it is also tangled up with the psychosocial atmosphere at work and job stress, making it difficult to satisfy end-user requirements. Further complications arise due to the cumulative or synergistic effects resulting from the interactions between them; thus it is necessary to systematically specify measureable limits.

Other aspects of interior design criteria are driven by the strategies for whole building: life safety, vertical transportation as well as some specialised services, and connectivity, the design criteria for the interfaces between the building and the outside world.

Building services engineers may also be involved with specifying the design criteria for systems for outdoor areas, such as external lighting and small power, water services (irrigation) etc. Although these are not 'indoor' areas, it is possible to define design criteria that can be used as a basis for design.

Each area within a building will have its own indoor design criteria. Building services engineers should not be left 'guessing' clients' requirements: clients have a responsibility to articulate how they want the building to work. Building services should confirm their understanding with respect to:

- occupancy patterns – numbers and duration
- the type of activity being undertaken – including duration
- known fixed equipment requiring mechanical or electrical connections, such as fume cupboards or drinks stations. A description of built-in equipment should include location, size and requirements for power, air, water, drainage and telecommunications connection
- allowances for future fixed equipment
- lighting
- small power and ICT outlets.

These criteria are often captured in room data sheets.

Thermal comfort

Thermal comfort perceived by a person depends on both environmental and personal factors. Environmental factors which can be controlled and hence can be specified are primarily defined in terms of air temperature, humidity, air supply rate and the quality of air. Personal factors include the level of clothing and activity level.

Industrial buildings are 'process' focused buildings and therefore 'comfort' is less relevant (apart from the small portion of the building that houses staff).

Table 6.5 shows a sample specification for thermal comfort for an office area – we can consider each aspect in turn.

Air temperature

Air temperature is a measure of the way people lose or gain heat from the space due to heat transfer mechanisms of radiation, convection and conduction. This is a dynamic exchange which is affected by the heat gains and losses in the locale, as illustrated in Figure 6.11.

Table 6.5 Sample design criteria: thermal comfort in an office.

Air temperature	Internal winter temperature: 21°C
	Internal summer temperature:* 26°C
Humidity	40–70%
Supply air rate	15 l/s/person
Air speed	0.5 m/s at 0.75 m AFFL
Air quality	VOC levels should not exceed 300 µg/m³ averaged over 8 hours
Pressure regime	Equalised with adjacent rooms

*The internal summer temperatures shall not be exceeded on average 3% or 9 days per year over 20 years.

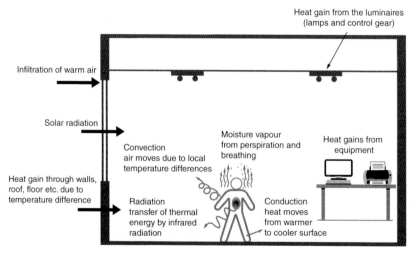

Figure 6.11 Heat gains to a space.

Relative humidity

Relative humidity is a measure of the water vapour content in the air. Low relative humidity levels can cause discomfort through drying of the mucous membranes and skin, and may also cause static electricity build-up and negatively affect the operations of some office equipment such as printers and computers. High relative humidity levels may lead to excessive perspiration, exacerbation of the effects of high temperature, feelings of 'closeness', etc. and the development of condensation on surfaces and within the interior of equipment and building structures which may develop mould and fungi.

Supply air rate

A sufficient supply of fresh air is required for the occupants and the functions in the rooms. This will:

- provide a continuous supply of oxygen necessary for human existence
- remove the products of respiration and occupation
- remove contaminants, such as new furniture and fittings which may emit volatile organic compounds, human bodies which give off odours and carbon dioxide, aerosols, gases, vapours, fumes (including cigarette smoke) and dust – some of which may be toxic, infectious, corrosive, flammable or otherwise hazardous
- enable dilution and control of airborne pathogenic material
- control excess humidity, arising from water vapour in the indoor air, to mitigate condensation
- remove heat from equipment
- supply make-up air, where local exhaust ventilation (LEV) is used, such as chemical fume cupboards
- provide air for fuel burning devices
- provide a means to control thermal comfort.

The volume of fresh air (make-up air) is determined by the size and use of the space to change the air in the room sufficiently. A variety of different units may be used including air changes per hour (ACH), litres/second per person (l/s/person), volume per hour (m^3/hr).

Also, the air speed should be specified at the relevant positions. Neither draught nor stagnant air is satisfactory.

Air quality

Indoor air quality is concerned with maintaining hygienic conditions for the health and welfare of occupants, or for processes such as electronic assembly and food preparation, protecting building finishes, fabrics, equipment and furnishings to reduce redecoration costs and protecting supply and extract air equipment or in the space itself from pollutants.

Indoor air quality is affected by pollutants originating indoors (from people, processes and the buildings fabric and fixtures) and external

pollutants entering spaces via the ventilation system, from adjacent rooms via doors and from the outside via gaps in the buildings fabric.
Pollutants originating indoors can be classified as:

■ chemical (e.g. NO_2, CO_2, carbon monoxide, radon, VOCs, body odours)
■ microbiological, (e.g. mould spores, bacteria)
■ particulate matter (e.g. smoke, dust, fibres).

As pollutants affect health, the minimum requirements are included in statutory legislation applicable to the particular space. These usually quantify upper levels in parts allowed for a period of time.
The building services systems can control the pollutants by a combination of ventilation rates, filtration and building airtightness.

Pressure regime

The pressure difference may either be balanced or positive or negative with respect to adjacent spaces as illustrated in Figure 6.12.
Where it is prudent to prevent the egress of air from one room to the next via the door (or other gaps) – for example due to toilet/kitchen smells or to avoid contamination – the air in the room must be at negative pressure relative to these areas and exhausted from within the room.

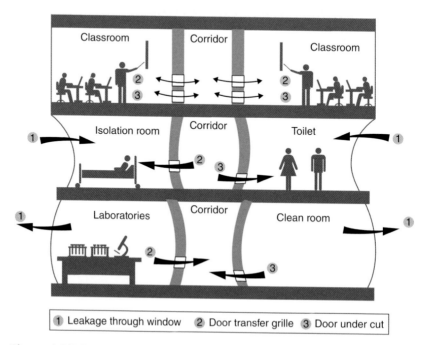

Figure 6.12 Pressure regimes in a building.

Negative pressure is created by balancing a room's ventilation system so that more air is mechanically exhausted from a room than is mechanically supplied. This creates a ventilation imbalance which the room ventilation makes up by continuously drawing in air from outside the room, via leakages in the facade, transfer grilles and around door frames. Pressure stabilisers may be provided, which operate in one direction only, allowing excess air to be directed to the area desired and assisting in maintaining room pressure differentials. When closed, they prevent significant reverse airflow.

Positive pressure is created by balancing a room's ventilation so that less air is mechanically exhausted from a room than is mechanically supplied. This creates a ventilation imbalance in which air, and any particles, are forced out of the room, keeping contaminants out for as long as the intake is properly filtered. This can be useful in an environment where nothing in the room is dangerous and the air needs to be kept as clean as possible, such as in a semiconductor assembly, clean room or laboratory where impurities can cause problems with the finished product.

A positive pressure regime is also used in firefighting staircases and lobbies to ensure that smoke does not enter the escape routes. There are various ways of configuring this – an example is shown in Figure 6.13.

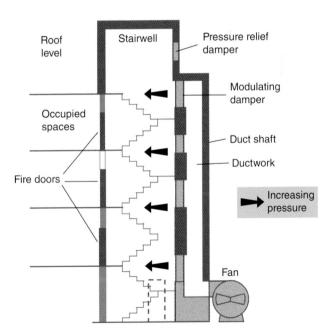

Figure 6.13 Stairwell pressurisation with dampers and multiple injection points to regulate pressure.

The reality of thermal comfort

Suppose, for example, we have a design internal temperature of 21°C. Theoretically, at least, this should be comfortable. But you don't get 21°C everywhere in a space. For a start you get vertical temperature gradients. Warm air rises. So we're going to get cool air at low level and warmer air at high level; and no matter what you do, you'll still get a vertical stratification through the space.

In addition, if you've got equipment or people, these will also be giving off heat and hence producing little convection currents. Then we might have a series of radiators each producing a rising column of hot air, plus if we've got cold windows in winter, then they will cool the air immediately adjacent to them which will fall – in other words, we're going to get a lot of air and other thermal movements in and around the space, which makes it difficult to ensure that we're achieving ideal conditions where we want them.

Now people vary in terms of their comfort, and there are many different factors that affect this, some of which you can perhaps adjust by clothing, by activities or by metabolism. For example, if you're running around you tend to generate more heat than if you're sitting still.

Now, in terms of the conditions that we are trying to control within the space – the various factors that affect all these heat interchanges – the air temperature of the space and the air velocity, which relate to the convective heat losses, and the surface temperatures which relate to the radiant heat losses, we also have one other factor: the amount of moisture in the air. And that's another way that we lose heat … by evaporation!

As we breathe out, and as we perspire, we experience a gradual loss of water from our bodies. It happens all the time. Of course, if it's really hot we may perspire more rapidly and noticeably, but it happens all the time, and the amount of heat lost by these processes is affected by the amount of water vapour in the air.

Therefore, when you're looking at thermal comfort factors, these are the things that you want to try to control with the actual systems.

Assuming that you're trying to create a reasonable air temperature, then that will depend on whether we're blowing warm or cool air into the space. Looking to control surface temperatures? Again this will depend on how hot or cold the space is. We can even change the average surface temperature by putting in hot objects such as radiators, or very cold objects such as cooled panels, which will affect the overall surface temperatures and which will affect the radiant interchange that we have with the space.

You can affect the air velocity by blowing air into the space at different speeds, thereby producing different distributions, and you can affect relative humidity by adding or removing water vapour from the air that's being supplied to the space.

There are also the physiological effects: perception of having 'control' contributes to a person feeling comfortable. All of these are things that can be controlled to some degree and will therefore affect comfort.

Finally, one thing that also affects our personal comfort is to have warm feet and a cool head. You don't want hot feet, but if you've got warm feet, that's comfortable, because cold feet are very uncomfortable, and you want a nice cool head for clear thinking. What this means is that you want the temperature to be a little lower at head level than it is at floor level – which is very difficult to achieve because, as we've seen, warm air rises! All we can do in reality is to try to ensure that the two are not too different, and we'll see how we do that with various systems a little later.

Table 6.6 Example design criteria for visual conditions.

Interior lighting
Daylighting: All habitable rooms will have an average daylight factor of at least 2%
Electric lighting: Illuminance 300–500 lux, at 750 mm AFFL
glare index 19
CRI 80
colour temperature 3500 K
maintenance factor 0.74
Controls: continuously dimmable with presence detectors
Emergency lighting: self-contained, non-maintained luminaires

Exterior lighting
Electric lighting:
Controls: external luminaires are controlled through a time switch, or daylight sensor, to prevent operation during daylight hours

Visual conditions

Visual design conditions for internal areas are defined in terms of daylighting, electrical lighting and the associated controls, for both the normal and emergency situations to achieve criteria in terms of lighting quality and energy efficiency. Table 6.6 is an example specification for the design criteria for visual conditions.

Daylighting

With reference to Figure 6.14, the amount of daylight reaching a particular spot inside a building in dependent upon:

- the sky component (SC), which is the light reaching the point directly from the sky
- the externally reflected component (ERC), which is the light that reaches the point after being reflected from surfaces outside the room such as buildings or roads
- the internally reflected component (IRC), which is the amount of light that reaches the point after being reflected from other surfaces in the room.

The amount of daylight outside: this constantly varies due to the season, time of day and prevalence of clouds. Hence, it is not possible to specify

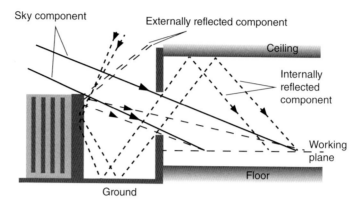

Figure 6.14 The three components of daylight.

the actual lux levels at a point inside a building. To avoid the difficulty of having to deal with frequent and severe fluctuations in the intensity of daylight, it is specified as a percentage of the outdoor luminance reaching a point indoors, assuming that the outdoor is a simplified overcast sky condition.

Daylighting is generally seen as desirable as it is 'natural', and windows also play a vital psychological role in providing a visual link with the outside world. However, daylight can also be a problem: daylight is not constant, it can cause glare and it disappears at night. Also, increasing the size of the window to let in more daylight increases the solar heat gain to the space and increases the noise penetration, both of which require mitigation measures. So daylighting design criteria have to be carefully blended with the artificial lighting to ensure a satisfactory balance to the room conditions.

Electric lighting

The illuminance or light level is the amount of light energy (luminous flux/lumens) reaching a defined surface area (the task level) per square metre. It is important to properly specify the task level, as this is where the light is required. Examples are given in Figure 6.15.

Glare can be generally divided into two types: discomfort glare and disability glare. Discomfort glare refers to the sensation experienced when the overall illumination is too bright – such as on looking out of a window on a very sunny day – and results in an instinctive desire to look away from the bright light source, or difficulty in seeing a task. Disability glare refers to reduced visibility of a target due to the presence of a light source elsewhere in the field, such as flashlights or visible lamp sources. The quantitative measure of glare is the glare index which recognises the visual conditions in which there is excessive contrast or an inappropriate distribution of light sources that disturbs the observer or limits the ability to distinguish details and objects.

Task level–
desk level and
monitor screen

Task level–
floor

Task level–
at the underside

Task level–
on the whiteboard

Figure 6.15 Examples of different task levels.

The colour rendering index (CRI) is used to compare lamps and quantify how good they are at reproducing colour. A set of test colours is reproduced by the lamp of interest, relative to how they are reproduced by an appropriate standard light source. A perfect colour rendering lamp would give a value of 100.

Some lamps provide good colour rendering properties and this may be necessary in areas where accurate colour appearance is important, such as for hairdressers, medical environments or art galleries. In which case the CRI may be 90 or even 100.

The colour temperature is the apparent colour of the light emitted by the lamp and is quantified by its correlated colour temperature (CCT). Most lamps produce some form of white light from cool to warm. This is their colour appearance. A warmer appearance is suitable for relaxed situations whereas a cooler appearance is used where high lighting levels are required and in work situations.

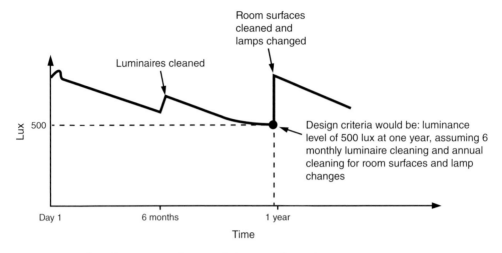

Figure 6.16 Effect of maintenance factor on lighting installation design.

Table 6.7 Design criteria parameters for lighting controls.

Specification parameter	Comments
Range of control	For example, On/Off, 0/25/50/100% stages, continuously variable
Inputs	For automatic controls, this should describe the parameter, for example:
	• automated timeclock
	• astronomnical timeclock(calculates sunrise and sunset times)
	• local and central control panels
	• handheld remote controls
	• daylight monitoring photocells
	• occupancy (presence or absence) detectors
	• serial data or volt-free contacts. e.g. alarm situation
Extent of localisation or centralisation	Centralised system for whole building

The maintenance factor accounts for the reduction in light levels over time. The illuminance from a lighting installation will vary from day one, according to the cleanliness of the luminaire and room surfaces and the strategy for failed lamp replacement: in any case, the light output from a lamp will degrade as it ages, as Figure 6.16 illustrates.

Controls

Lighting control systems can be used to reduce energy costs and extend the life of lamps and ballasts. It can also be used to create moods and ambience by setting difference scenes. Table 6.7 illustrates the areas needed to be covered in the design criteria for lighting controls for any space or zone.

Emergency lighting

Emergency lighting is used to provide some lighting if there is a failure in the normal electrical supply to all or part of a building. The local enforcing authorities set the requirements for the emergency escape lighting to provide an illuminated route to direct people out of the building along the safest (and normally shortest) route. The route will be identified by means of directional signage (green and white signs) and, outside, the congregation point or open area must be lit.

In some circumstances, it may not be appropriate for some, or all, of the occupants to leave the building, and additional standby lighting is provided. Situations could be:

- in a medical environment where a treatment is being performed
- in plant rooms where maintenance may be being carried out and it could be unsafe to leave part way through a procedure
- in a catering facility where, if the lighting failed, it would be safer for the cooks to make sure that all the catering equipment is turned off and made safe before leaving the kitchen
- in a financial organisation such as a dealing room where there would be financial penalties.

The reality of designing for visual comfort

Trying to design comfortable lighting is not just about installing enough lamps, it's also thinking about how we see. Unfortunately this also introduces a whole new set of specialist terms, some of which are unique to the lighting industry: for example we use lamps not bulbs – bulbs grow in the garden – luminaires not light fittings and a host of other specialist terms such as: luminance, illuminance, reflectance and, daylight factor.

But the biggest problem with lighting is that we don't actually design for what we see!

Humans, like most animals, have evolved to be able to see at a wide range of lighting levels, and those that couldn't adapt never made it down the evolutionary chain. This means that we can adjust our visual sensitivity to a wide range of illumination levels – which makes it difficult to specify a required lighting level.

All we can hope to do is to follow the guidelines of our peers by way of the relevant lighting code of practice, in which the recommended levels have taken into account such factors as:

- enough light to see and work safely
- enough light to make us feel comfortable
- enough light to take into account ageing of the lamps and general cleanliness of the room surfaces etc.

Glare and colour also play an important role. It is not sufficient, for example, to ensure that the lighting is not blindingly glaring; we need to ensure much lower levels of luminaire brightness so that users of the space do not become too tired or uncomfortable by the end of the day – the concept of 'discomfort glare'.

Similarly, we are influenced both by the colour appearance of the light source and its abilities to render or display accurately, any colours seen under them. For example, in the northern temperate zones people prefer lamps with a warm colour tinge, thereby making white room surfaces seem warm and spaces invitingly cosy, while people living under tropical conditions would not thank you for anything less than a colder looking lamp.

And while we're on the subject, generally those light sources that accurately display colours seen under them are either slightly more expensive or produce a slightly lower output than so-called 'standard' lamps which have acceptable colour rendering properties. And so if you decide that you must use these more accurate lamps, then you'd better inform the client's maintenance team or else at the first lamp change a cheaper alternative might be substituted.

Daylight can also be a problem. What?! 'But daylight is free!' I hear you cry. Oh no it isn't! Increase the size of the window to let in more daylight and you also increase the solar heat gain to the space and increase the noise penetration, both of which require expensive solutions to control. So daylight is not free, it has to be carefully blended with the artificial lighting to ensure a satisfactory balance to the room conditions. And of course daylight varies both during the day and with the seasons.

Exterior lighting is more often defined with regard to placement, intensity, timing, duration and colour, according to their function, for example:

■ street lighting within the residential built development areas
■ highways lighting on the primary access points and primary internal roads
■ security and health and safety lighting at the local community centre or local school
■ residents are likely to install security lighting at their properties.

Acoustic conditions

The acoustic performance of a space is defined in terms of noise levels, degree of privacy and speech intelligibility. These are dependent upon the make-up of the walls, floors, ceilings and openings (intentional or otherwise) and the adjacencies to other areas. These are usually specified by architects. Building services engineers will be responsible for ensuring that the noise and vibration contribution from the building services equipment maintain the defined conditions.

Design criteria for public address systems, fire alarm sounders and audiovisual systems tend to focus on intelligibility and well as audibility.

Electromagnetic and electrostatic environment

Electrical and magnetic fields are emitted by electrical distribution equipment and appliances. These electromagnetic emissions can interfere with the normal operation of other devices or systems. With respect to building services, potential sources of electromagnetic interference include electric motors, transformers, lamp ballasts, power supplies and cables carrying electricity, particularly high voltage. Potential susceptible recipients include low voltage electrical and signal cables and any electrical appliances. The problems range from minor annoyances to complete failure of equipment and systems.

Electromagnetic interference

Most car drivers accept that when they drive near electricity pylons, their radio listening pleasure may be interrupted by loud crackles and/or buzzing noises but with the increased use of electronic equipment, the problem of interference has become one of our prime concerns.

Although most forms of interference are usually tolerated as being 'just one of those things that you cannot do much about', the design of sophisticated modern equipment has become so susceptible to electromagnetic interference that some form of regulation has had to be agreed.

Electromagnetic compatibility (EMC) describes the ability of electronic and electrical systems or components to work correctly when they are close together. In practice this means that the electromagnetic disturbances from each item of equipment must be limited and also that each item must have an adequate level of immunity to the disturbances in its environment.

In terms of specified design criteria, building services engineers should recognise the need to ensure that both appliance emissions and susceptibility or immunity are within recognised regulations.

Life safety criteria

The requirements for life safety services – for example, emergency escape lighting, fire alarm systems, installations for fire pumps, fire rescue service lifts, smoke and heat extraction equipment – need to be defined.

Firefighting shafts, as illustrated in Figure 6.17, are provided in larger buildings to help firefighters reach floors further away from the building's access points. They enable firefighting operations to start quickly, and in comparative safety, by providing a safe route from the point of entry to the floor where the fire has occurred.

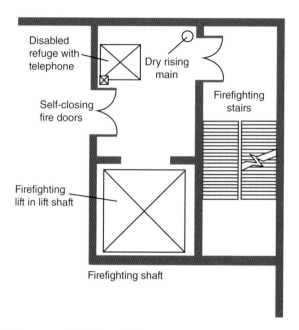

Figure 6.17 Elements of firefighting shafts.

Table 6.8 Sample design criteria: vertical transportation in an office.

Lifts to handle highest demand with response time as indicated by hall call registration not exceeding an average of 30 seconds during 30 minutes of heavy two-way traffic. Minimum handling capacity for 12% of office population arriving within five minutes of each other.

Vertical transportation

Vertical transportation is concerned with the means of people and goods travelling between floors in a building. All buildings with more than one storey must obviously have at least one set of stairs, and the provision of stairs is a very important consideration when designing buildings, in order to ensure that all the occupants of the building can escape safely in the event of a fire.

A lift traffic analysis identifies the size, speed and capacity of lifts needed, to provide the levels of service required. Table 6.8 provides an example specification.

The lift traffic analysis is based on assumptions about the flow of the building population, such as when/where do they enter/leave and are there facilities, such as restaurants or gyms, in the buildings that have a strong effect on the usual passenger flow.

The more accurately the passenger movement can be modelled, the more accurate will be the results of the traffic analysis. Once these are modelled, an accurate calculation of the lift performance can be made.

The outputs of a traffic analysis will give an accurate indication of the quality and quantity of lift service provided, that is, what the waiting times are and what percentage of the building population can be transported by the lifts in a given time period.

Specialist services

Some more specialised buildings will require further design criteria. These may include

■ vacuum, compressed air and medical gases for healthcare buildings
■ industrial applications.

Connectivity

Connectivity refers to the interfaces between the building and the outside world. These are generally the utility services necessary for the occupants to live and carry out activities inside the buildings, and without which the building would be unusable. These include electricity, gas, water and sewage services; information services such as telephone and television cable services; and district heating and cooling systems. On a campus type site, other utility services such as medical gases, pneumatic tubes, wide and local area network connections (WANs and LANs) may also be needed. Building services engineers should check the availability of such services because non-availability will inform design decisions; for example, the non-availability of gas (say in a remote area) would preclude the use of a gas-based kitchen – unless bottled gas is provided.

Controlled outdoor environment

While building services engineers mainly focus on the environment inside a building, there are areas outside the building that may also be utilised by building users, for which building services engineers will need to specify design criteria.

Domestic water

External water may be required for fixed appliances outside the building such as fountains, water features and swimming pools. It may also be necessary to provide taps for operational use such as plugging in garden or cleaning equipment. Potable water may be required for drinking purposes via taps or drinks fountains: the design criteria should specify the locations.

Irrigation water

Irrigation is the act of artificially applying water to soil to maintain desirable conditions, such as for plant growth or for sports fields (including golf courses). The design criteria should specify the irrigation zones, according to areas or types of planting and soil conditions, and assumptions on climatic conditions. They should identify the properties of the water, including the state (liquid or gaseous) and quality, including any chemical additives.

External lighting

Exterior area lighting associated with a building project covers a range of different situations – such as lighting for car parks, facades, sports areas, landscaped areas, water features, artwork, signage and private roads/footpaths/cycle paths – these are more often defined with regard to placement, intensity, timing, duration and colour. Design criteria for exterior lighting and signage for public roads, cycle paths, footpaths and amenities (such as bus stops) are controlled by central and local government and are usually required to be designed to adoptable standards: installed in accordance with the standards of the entity who will take over and maintain the installation for its lifetime.

Cooling systems

Outdoor spaces may benefit from mechanical cooling systems to make them more usable; for example, areas for sporting spectators. The design criteria should specify the areas to be covered and the target temperature and humidity.

Heating systems

Outdoor spaces may benefit from mechanical heating systems to make them more usable, for example restaurant areas. The design criteria should specify the areas to be covered and the target temperature and humidity.

Pesticide control systems

Systems may be provided to spray pesticides in a fine mist to kill mosquitoes and other insects outdoors. The design criteria would specify the types of pesticide used, the timing and the method of activation: wind, rain and motion sensors may be used to reduce unnecessary pesticide usage and inadvertent application when people or pets are present. Figure 6.18 illustrates a possible installation arrangement.

External power

External power may be required for fixed appliances outside the building, such as access control equipment, CCTV cameras and vehicle barriers, or for building services equipment such as pumps. It may also be necessary to provide socket outlets for operational use, such as plugging in garden or cleaning equipment, or general-purpose power.

Figure 6.18 Pesticide control system integrated into a fence.

Building services engineers may also be involved with developing the architectural form of a building so as to pre-eliminate microclimatic challenges; for example, if well designed, courtyards can give a degree of protection from wind-borne dust entering a building.

6.3   Voluntary codes and practices

While legal and regulatory instruments will set minimum standards which will continue to be revised, added to and improved upon, they will almost inevitably lag behind technological advances, accepted knowledge or areas of increasing concern or action. Hence, voluntary codes and practices may be included in the design criteria.

The emergence and evolution of voluntary building environmental assessment tools is having a significant impact on building services engineering design. Particularly as they necessitate a more thoughtful, multidisciplinary and integrated approach to design, with all parties involved from the start of the process.

Voluntary codes are non-statutory and often self-regulatory measures. The building services engineer should be aware of what schemes are available and suitable for a particular project, but it would usually be the choice of the client to subscribe to any particular scheme.

Voluntary codes and practices can be categorised as incentive based, eco-labelling or benchmarking schemes.

Incentive schemes

An incentive scheme encourages specific actions or behaviours, which are rewarded. These are usually run by governments and are thus subject to changes and removal as government policy changes, so this book does not attempt to describe current incentive schemes.

Possible incentive schemes

Renewable electricity generating technology
If end-users install technology, such as solar PV, wind turbine, micro-CHP or micro-hydro, they may be eligible for a feed-in tariff (FIT) scheme which pays generators a tariff for each unit of electricity generated and an additional tariff for each unit of electricity exported back to grid.

Energy conservation measures
Reduced VAT or tax rebates may be offered on energy saving materials such as insulation and heating controls.

Energy efficient equipment
Enhanced capital allowances (ECAs) may be provided, which allow an offset against tax for capital investment in technologies that are classified as leading-edge energy efficiency equipment.

Offsetting
There may be relaxing planning conditions or increased approval times given in return for a tangible benefit such as the provision of services, for example amenity lighting, for the benefit of the community.

Eco-labelling

In eco-labelling schemes, typically, a building or specific items within a building are assessed with reference to a set of quantitative and qualitative performance criteria and are awarded credits when in compliance with the various aspects that the assessment covers. With respect to building services these tend to be energy assessment or energy labelling schemes.

Typically the schemes are run by public authorities, environmental agencies or trade associations. Participants can use their participation to gain certification for a beneficial purpose, such as image building or complying with corporate social responsibility targets.

Examples of possible eco-labelling schemes

Equipment certification

Particular items such as domestic appliances, lamps and air-conditioning units may be labelled according to energy consumption, noise and disposability factors. Windows may be labelled according to emissivity (the relative ability of its surface to emit energy by radiation) and solar control ability on energy and CO_2 savings.

Whole building certification

This is a certificate which considers such elements as building envelope, windows, heating, electrical and ventilation installations, lighting, heat sources (including boilers, CHP units) cooling systems and others. The certification may also include advice and information on how to improve energy performance.

Summary

It would be exorbitant to design a building to cope with all the possible forms of unusual, severe or unseasonal weather, such as extreme heat or cold snaps, so it is important to establish the range of criteria for external conditions and to ensure that the end-users and clients understand the impact if the external criteria are out of range. A range of internal design conditions also needs to be established. These are based on minimum statutory requirements (where they exist) and enhanced where required by the client or by recommendation of building services engineers. The combination of the external and internal design conditions sets the parameters for the performance of the building services engineering systems.

Reference

CIBSE (2005), AM10 – *Natural Ventilation in Non-Domestic Buildings*, London, UK: CIBSE.

7 System descriptions

Building Services Design Management, First Edition. Jackie Portman.
© 2014 John Wiley & Sons, Ltd. Published 2014 by John Wiley & Sons, Ltd.

The building services solution for a building will comprise a number of systems consisting of a collection of components that are organised for a particular purpose. Some elements of each of the building services systems will interface with other building services systems and/or the building fabric. Although there are very many ways of configuring a system, there is often a common logic to the arrangement.

For each building services system there are key design parameters which need to be defined by building services engineers to ensure that the required interior design criteria are achieved, bearing in mind any relevant external design criteria. Care should be taken to ensure that these are coordinated with the design criteria for the architecture and civil and structural engineering systems as well as the budget and programme.

This chapter describes the principles of the main building services systems. It does not provide guidance on design methods or options. There are plenty of other books which cover these. The aim is to provide a strategic design management overview.

Preference engineering

For when we talk about establishing satisfactory internal conditions for a building, things aren't terribly precise. Whereas with structural engineering you might have a choice, within reason, of only concrete or steel, if you hand a design to ten different structural engineers you'll end up with virtually the same size beams and the same factors of safety. However, give a specification to ten different building services engineers and you could end up with ten completely different designs. Some might go for a displacement ventilation system, some might go for a fan coil system and others might choose a mixed mode system.

In other words, there are many different choices and different degrees of implementation. And together with the choices of different equipment, different manufacturers, different sizes, different supply temperatures and so on, you could end up with a choice of a number of completely different systems, let alone a number of completely different design processes, all of which are compliant with the requirements of the brief.

7.1 Public utility services connections

A public utility as an organisation serving a premises with electricity, gas, water (potable water, irrigation water, fire service water, sewage and storm water), ICT services (broadband television, fixed telephony and internet services) and broadcast-communications services (radio and television). Public utility services might also include the likes of urban combined heat and power (CHP) schemes and district heating and cooling schemes. Following deregulation, the 'public' utility services are often privately owned, but the term 'public utility' tends to differentiate it from 'private utility' services which are those produced by a building for its own use, such as local diesel generators, solar power and wind turbines for electricity and boreholes for water.

Building services engineers are responsible for determining the availability, quality, reliability, capacity and supply conditions of utility services required from the public utility services provider to meet the needs of the building. The public utility services providers determine how this will be provided and the apparatus and equipment and demarcation point; the building services engineers will then need to coordinate their requirements with the particulars of the building. This includes the provision of space, rooms and distribution routes, together with any specific access requirements and requirements for finishes and building services systems within the spaces. Utility services may be buried in the ground, fixed to buildings and other structures, suspended in the air or distributed wirelessly.

It is important to realistically assess the maximum demand, as this directly impacts the costs associated with having both the physical space and the capacity available. This can be represented as a load profile – a two-dimensional chart showing the instantaneous load over time – which represents a convenient way to visualise how the system loads change with time. Some examples are included in Figure 7.1.

This is achieved by estimating the energy demand. Building services engineers will need to understand the patterns of occupancy – hours and days during different seasons and for different scenarios of

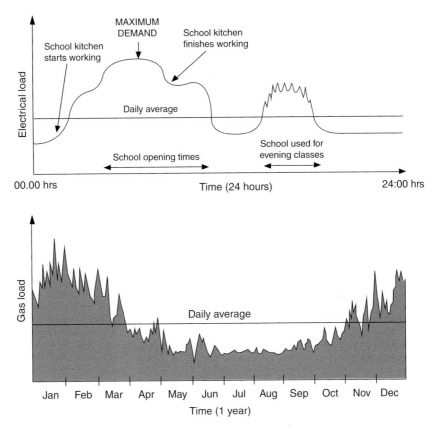

Figure 7.1 Electrical and gas load profiles for a school building (note the different timescales).

use within the building – in order to make realistic estimates of interrelated diversity. This, in some ways, is the hardest part of the design process and it is a brave building services engineer who can state categorically, at the start of a design, what the eventual load profile will be on the finished building. To assist in this initial assessment, it is important to gather information about the building and its potential use.

All buildings are different and no two buildings will have the same utility services usage, even if they are of the same type and built for the same purpose. For example, they may have different constructional parameters or be built in another part of the world.

Electricity

Figure 7.2 describes the constituent parts of a specification for the incoming electrical supply, and Figure 7.3 illustrates examples for a range of buildings.

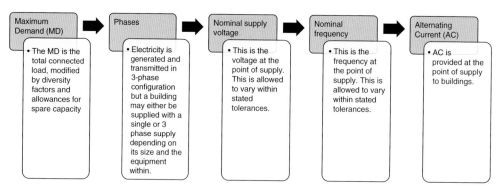

Figure 7.2 Constituent parts of specification for incoming electrical supply.

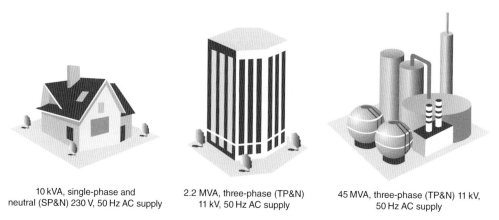

10 kVA, single-phase and neutral (SP&N) 230 V, 50 Hz AC supply   2.2 MVA, three-phase (TP&N) 11 kV, 50 Hz AC supply   45 MVA, three-phase (TP&N) 11 kV, 50 Hz AC supply

Figure 7.3 Example specifications for electrical supplies.

Gas

Figure 7.4 describes the constituent parts of a specification for the incoming gas supply. Figure 7.5 illustrates examples for a range of buildings.

Again, building services engineers will need to estimate the load profile; with gas, the maximum demand is usually driven by the catering requirements.

Water

Water services comprise supply water (potable, irrigation and fire water), and wastewater (sewage and storm water). All of which will be specified separately. Table 7.1 illustrates examples for a range of buildings.

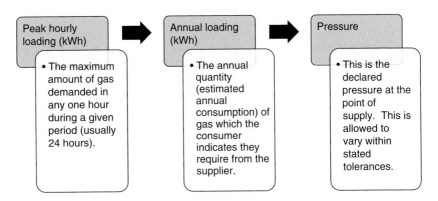

Figure 7.4 Constituent parts of specification for incoming gas supply.

25 kWh peak gas load and 15 MWh annual average usage at low pressure

1.55 MWh peak gas load and 800 MWh annual average usage at medium pressure

5 MWh peak gas load and 60,000 MWh annual average usage at high pressure

Figure 7.5 Example specifications for gas supplies.

Information and broadcast communications

Information and broadcast communications services can enter a building by cable or via airwaves. Airwaves would either be direct to a device or intercepted by a receiving device for onward transmission either via a cable system or by local wireless transmission.

Owing to the fast-moving nature of the provision of information and broadcast communication services, it may not be sensible to finalise requirements until nearer the completion of a project. Thus, it may be prudent to either allow space only or minimal cable containment during the earlier design stages, and design the details of the cables and supporting apparatus later, to suit the current technologies and services available.

Table 7.1 Examples of water supply specifications.

Potable cold water	25 mm diameter water supply pipes	2 No 90 mm diameter water supply pipes	4 No 300 mm diameter water supply pipes
Irrigation/ process water	N/A	35 mm diameter irrigation water supply from borehole	90 mm process water supply from local river
Fire system water	N/A	150 mm diameter fire water supply at 5 bar pressure	150 mm diameter fire water ring supply at 15 bar pressure
Sewage water	10 mm diameter sewage discharge pipe	150 mm diameter sewage discharge pipe	300 mm diameter sewage discharge pipe
Storm water	N/A	100 mm diameter stormwater pipe, to discharge to local river	1 No 100 mm diameter storm water pipe to connect to mains system + 1 No 100 mm diameter storm water pipe to discharge to local reservoir

7.2   Ventilation

A ventilation system provides fresh air to allow people and processes to function. A ventilation system comprises a source of fresh air and a means of distributing it to the required space. A reverse process is required to remove the stale air. The ventilation system may also form part or all of the heating or cooling system. Ventilation systems range from natural ventilation to full mechanical ventilation – or hybrids, being a mixture of the two.

For a naturally ventilated system the sources of fresh air are either intentional displacement though specified openings or unintentional via infiltration. Similarly, exfiltration, the unintentional leakage of air from the inside to the outside, due to a higher pressure may occur, as illustrated in Figure 7.6.

Air movement is based on the stack effect and relies on wind power and temperature difference to cause a pressure difference as illustrated in Figure 7.7.

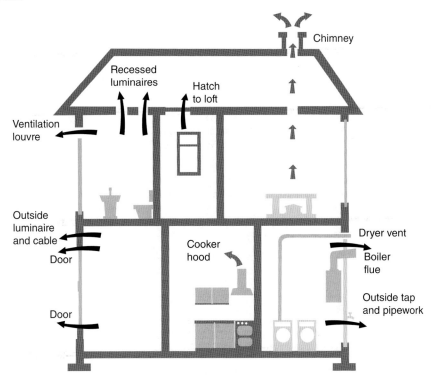

Figure 7.6 Sources of fresh air and means of removal.

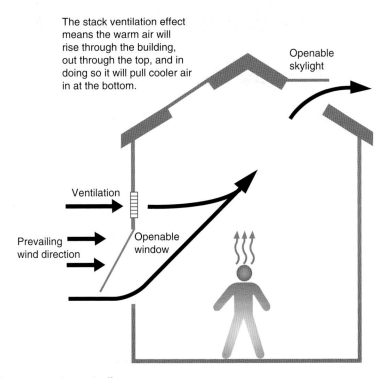

Figure 7.7 The stack effect.

The stack effect can be further developed with specially designed features incorporated to assist airflow by heating part of the building fabric by solar radiation, resulting in a greater temperature difference, hence larger airflow, as illustrated in Figure 7.8.

A natural ventilation system is limited by the following factors:

■ the actual wind direction may vary from the prevailing
■ the wind strength may vary
■ the outside temperature may drop below a level sufficient to produce the stack effect
■ the space planning may not permit adequate ceiling height – a deep-plan building, where some rooms are a long way away from any perimeter windows; a distance of 7 m is usually considered the practical maximum
■ in certain areas, such as kitchens and toilets, it may never be possible for a natural ventilation system to provide sufficient fresh air.

It may not be possible to rely upon natural ventilation, and some mechanical ventilation may have to be included. For mechanically ventilated systems the sources of fresh air are intentionally drawn in via bespoke equipment, and the air is distributed from the source to terminal devices via a system of ductwork. The system is designed to deliver the required volume of air to a space at the required temperature and velocity, and acceptable sound level, all within the space available. Thermal insulation is applied to ductwork to reduce heat exchange and to prevent condensation. Air moves through the ductwork in response to pressure differences created by fans. Ductwork comprises a system of main and branch ducts, with plenums fitted with vibration isolators, volume control dampers and smoke/fire dampers, as well as supports and hangers as illustrated in Figure 7.9. The fresh air provision for different areas can be adjusted by means of manual or automatic dampers or constant air volume devices.

The terminal devices can be diffusers (which discharge a supply of air in a spreading pattern), grilles (louvred cover for an opening through which air passes) or air terminal units. These may be located in the ceiling, walls or floors. The relative locations of supply and extract terminals and their design air volume rates will determine the basic airflow between adjacent spaces.

Cleaning and access doors are required to give access to plant items and ductwork components for inspection, maintenance, cleaning and replacement. They must be of sufficient size to permit safe access for the required functions.

Another type of ventilation is the local exhaust ventilation (LEV) system, provided in science laboratories (in the form of fume cupboards) and other spaces to remove noxious fumes and gas discharges and to

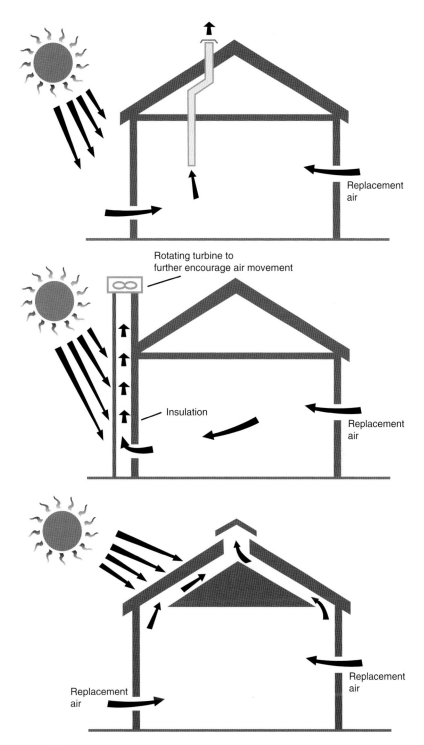

Figure 7.8 Natural ventilation principles.

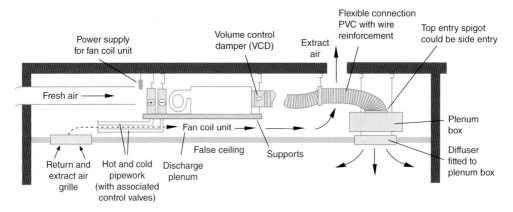

Figure 7.9 Components of a typical ventilation system.

extract dust. Simple LEV systems comprise a receptor or capture hood, extract ductwork and a local fan.

7.3   Heating

A heating system contributes to the thermal comfort in particular space when otherwise the heat loss would be unacceptable. A heating system comprises a heat source, distribution system and heat emitters.

The primary energy for the heat source may be a single or a combination of utility service suppliers, district heating systems or in-situ supplies (e.g. a diesel generator), sustainable sources or recycled 'waste' energy (e.g. biomass fuel or used cooking oil). If fuel sources are to be stored on site, the design needs to consider availability of materials, delivery (access for fuel delivery vehicles), storage (space and conditions) and transport of materials to the point of use.

An energy exchanger converts these primary sources to energy stored in the heat transfer medium (water or steam). There may be waste products from this process.

The distribution system is the equipment which conducts the heating medium which transfers the heat energy from the source to the heat emitters.

Heat emitters transfer the energy from the heating medium to the air, at the appropriate temperature, rate and location. Examples are radiators, natural and fan convectors, radiant panels and underfloor heating. Radiant panels use the principles of radiant heat to transfer energy from an emitting heat source, rather than transferring the heat by convection heating (as used in conventional radiators).

Positioning of heat emitters is important in achieving optimum performance. The standard approach is to position radiators below windows, as this is the area of maximum heat loss from a building. This

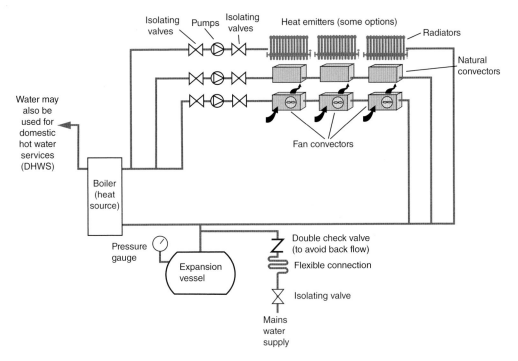

Figure 7.10 Basic components of a simple heating distribution system.

also helps to prevent condensation forming on cold windows. As insulation standards improve with the use of thermally efficient double or even triple glazing, this is less of a consideration, so positioning needs to be considered in relation to the thermal insulation of each building.

Greater flexibility in positioning of radiators is achieved through the use of slim-line radiators that protrude only a short distance into the room or corridor. Low water content radiators are particularly compact and therefore more versatile from a positioning perspective. In some cases, depending on the nature of the building fabric, it may prove beneficial to use built-in radiators, recessed into walls so that there is no intrusion into the heated space, providing this does not compromise their heating capability. Figure 7.10 outlines the principles of a very basic heating system.

The size of radiator is determined by the volume of the room, the desired temperature and the level of insulation of the room. A radiator that is too small for the space will not achieve the required temperature on cold days. One that is too big will achieve the required temperature more quickly but could waste energy in doing so if the thermostatic temperature controls do not react quickly enough, or if the radiator is a high mass unit which reacts slowly.

The associated control systems manage the space conditions by modifying input energy, temperatures and flow rates of the heating medium. Heating systems may provide local control where required, using thermostatic radiator valves (TRVs), but these are subject to tampering and it may be necessary to arrange them so that they are accessible only to designated operations staff.

Underfloor heating systems use fluid flowing in pipes (hydronic systems) or electrical heating elements, such as cables or carbon mesh, to heat the area above the floor by means of conduction (feet on floor), radiation and convection (as the heated surface influences the density of the air above). Hydronic systems use water as the heat transfer fluid in a closed loop that recirculates between the floor and the heat source. Underfloor heating can be used beneath a range of floor surfaces, including traditionally cold materials such as stone and tiles.

7.4   Cooling

A cooling system contributes to the thermal comfort in space when otherwise the heat gain would be unacceptable. In temperate climates specific cooling systems may not always be provided as cooling requirements are satisfied as part of the ventilation system or part of an air-conditioning system. There are various methods for cooling buildings:

- Night-time cooling is where cool air is passed through the building at night. This may be achieved by opening windows and using a purely natural system. Most buildings however, require mechanical ventilation to obtain the high level of air change to cool down the building fabric. This method is particularly suited to areas where the summer temperature drops off at night-time.
- Evaporative cooling is where water is sprayed into an air stream, and the water evaporates, taking away the heat from the space and thus causing a drop in the temperature.
- Desiccant cooling is where air from the outside is dehumidified using desiccant salts or mechanical dehumidifiers before entering the space. Decreasing the humidity allows people to sweat more and hence cool themselves.
- Ice storage is used to provide chilled water for terminal units. The advantage of using ice is that it acts as a buffer: in ice storage systems, a refrigeration plant generates an ice bank during off-peak periods, and this is then melted to provide chilled water for use during peak periods.
- Chilled beams and ceilings – chilled water circulates through cooling units, called beams, or through pipes incorporated into a ceiling. As warm air rises to the ceiling and touches the cool beam it is cooled and descends into the room.
- Hollow floor slabs can be filled with chilled air or night time cool air, so that the concrete mass has a reduced temperature. The coolth is subsequently released into the space.

- Ground source cooling which makes use of naturally cooler subsoil. A series of air pipes are laid in the ground at a suitable depth, and as air is forced through the pipes it is cooled.
- Sea water/river/lake/aquifer water cooling relies on piping in cool water, either for direct circulation to terminal devices or via a heat exchanger.

However, while these methods may be viable when cooling is required for people, they do not constitute air-conditioning and may not be suitable when cooling is required for processes.

7.5   Air-conditioning

An air-conditioning system modifies the properties of air, primarily temperature and humidity – but may also include filtration – to provide favourable conditions. This is for the benefit of the building's occupants and/or to suit the processes in the building. The cooling component is based on an estimation of the maximum occupancy and equipment heat generation.

Nomenclature clarified

Air-conditioning – full air-conditioning is full control of the air condition – that's freshness, temperature, humidity, cleanliness – the works. If you're not controlling humidity then theoretically it's called comfort cooling, but this is something of a lost cause today because the term air-conditioning is used for everything today. You have cars with air-conditioning, houses with air-conditioning, trains with air-conditioning, but in most cases they just mean cooling. But strictly speaking, full air-conditioning in its widest sense, is the control of all the aspects of the condition of the air including the amount of water vapour present in it.

There are two main arrangements used: localised (treating air locally to the point of supply) and centralised (treating air at a common point and distributing to the points of supply).

Split air-conditioners have two main parts: the outdoor unit is the section which generates the cold refrigerant gas and the indoor unit uses this cold refrigerant to cool the air in a space.

In a centralised system the fresh air, sources of heating and cooling, air filters, humidifying and dehumidifying processes are all combined in a single place – the air-handling unit (AHU). These are widely used as a package unit which incorporates all the main plant items, as illustrated in Figure 7.11.

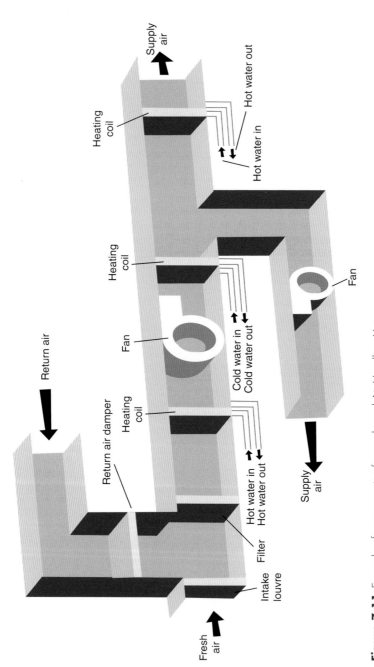

Figure 7.11 Example of components of a packaged Air Handling Unit.

Pipework, ductwork and electrical connections are made after the unit is set in place on site. The ductwork provides the distribution system for the treated transmission air. Other devices such as fans and dampers are added. The final termination units to emit the treated air into the space. There may be recirculation of used air and heat recovery.

By introducing fresh air into the system, the requirements for ventilation, for part or all of the building, may also be satisfied.

7.6   Water systems

Supply water systems are provided for potable water, irrigation water and fire water.

■ Potable water is water safe enough to be consumed by humans or used with low risk of immediate or long-term harm. It is used for drinking, washing, cooking, cleaning and commercial and industrial uses. Both hot and cold water need to be considered: they are usually referred to as domestic hot and cold water services respectively.

■ Irrigation water which is used primarily in the production of agricultural crops or for livestock, but it may also include garden watering.

■ Fire water which is used to for extinguishing fires.

Although these will be standalone systems they will involve sourcing supplies, treating it (if necessary), providing storage and distributing within a building to the point of supply. The design needs to be such that it arrives at the final outlet at the correct temperature and pressure.

Water removal services are provided to remove wastewater and storm water from buildings.

■ Wastewater systems are provide to remove soil water from toilets and wastewater from basins, baths, showers, and so on.

■ Surface water systems (also sometimes called storm water systems) are provided to remove naturally occurring water due to rainfall.

Hot and cold domestic water services

The sources for cold water supplies may comprise a single or a combination of public utility services, a temporary supply (tanker or bowser delivery), campus/site-wide sources, in-situ supplies (e.g. a well, borehole, river or water course), sustainable sources, including recycled (e.g. from rainwater, grey water or treated sewage effluent).

The hot water services derive from the cold water services, with the heating being provided by oil- or gas-fired boiler, electric immersion heater, solar water heating, heat exchangers working with site-wide distribution systems or heat pumps, which transfer heat energy from a heat source to a heat sink against a temperature gradient. Hot water generation may either be instantaneous or stored ready to be drawn off when required.

Water storage is provided to meet the ongoing demand if the mains supply is cut off for any reason. The quantity will depend upon the use to which the water is put and the number of consumers. In older buildings, it was common for gravity-operated storage tanks to be located on the top floor of the building. Nowadays it is more common to use pressurised tanks, connected to the supply pumps, and located at lower levels.

The distribution system consists of pipework to carry the water, pumps to maintain adequate pressure for movement and valves to allow for isolating parts of the system. Important issues that affect the design of the distribution are:

■ Materials used – which will depend on the climate, immediate location (above ground or underground) and types of fluids being transported, and also the required life expectancy.
■ Managing water pressure – a certain water pressure is required to ensure that water is pushed through pipes. Different end devices require different pressures. Valves and other devices can be installed to reduce water pressure; in other cases pumps are introduced to increase pressure.
■ Preventing backflow – this is the unwanted reversal of the flow of water from its intended direction in the pipework distribution system. Backflow is dangerous because it can lead to cross-contamination, for example allowing drinking water in pipe systems to become polluted and unusable. The most basic means of preventing backflow is an air gap, as illustrated in Figure 7.12.

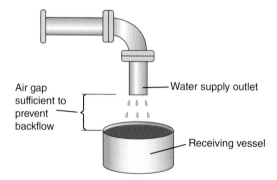

Figure 7.12 Air gap, preventing backflow.

An air gap is a vertical, physical separation between the end of a water supply outlet and the flood-level rim of a receiving vessel. This separation must be at least twice the diameter of the water supply outlet and not less than 2–3 cm. An air gap is considered the maximum protection available against backpressure backflow or back siphonage but it is not always practical and can easily be bypassed. Alternatively a mechanical backflow preventer, which provides a physical barrier to backflow can be installed.

■ Avoiding dead legs – these occur in hot water systems, and refers to where water does not move for a period of time. The most common time for dead legs to occur is at night when hot water is not used and the contents of the pipes and appliances cools down. When the hot water outlets are turned on the following morning, the cooler water is drawn off before hot water reaches the outlet. This could take some time if long runs of pipework are involved. Also when water is stored at around 20–45 °C it becomes more susceptible to bacteria growth, and overnight gives adequate time for bacteria to multiply. This can happen even if the pipework is insulated.

The quality of water, and hence its treatment requirements, varies according to the nature of the source. Therefore, there is not one standard system of water treatment; each water system will have its own requirements which may include:

■ disinfection to kill off harmful organisms in the water so that infection by disease will not occur when the water is used for domestic purposes
■ softening to improve the effectiveness of cleaning products and soaps and for increasing the lifespan of clothing, household appliances and pipes and pumps that form part of the plumbing system
■ removal of contaminants, such as suspended solids, bacteria, algae, viruses, fungi, minerals such as iron, manganese and sulphur, and other chemical pollutants such as fertilisers
■ addition of fluoride
■ taste and odour control
■ acidity/alkalinity correction.

Storage is required for both hot and cold water. This is sized to be able to service the demand in case the source of supply fails.

There are very many ways that domestic hot and cold water systems can be configured.

Alternatively, a number of localised hot water distribution systems could be designed, whereby hot water heaters are located closer to points of use, served from smaller local plant room. It is easier to maintain hot water distribution temperatures within recommended values. Balancing water flow rates in the hot water secondary distribution system becomes less of a problem, and distribution losses are reduced. A small, localised hot water distribution system may use a gas-fired water heater or an electric heater.

Irrigation systems

Irrigation water does not need to be drinking water quality. Potential sources include rainwater, river water and groundwater (from deep boreholes or shallow depth). Sustainable water management solutions involve abstracting or capturing water, storage and pumping to the points of use. Water treatment, to suit the final use will usually be required, and it should be routinely tested for water quality.

Fire water systems

A key design principle is to ensure that firefighters have immediate access to adequate supplies of water. There are several types of fire systems:

■ fire hydrants
■ wet and dry risers
■ automatic fire sprinkler systems.

A fire hydrant system consists of an assembly of pipework and components that allow firefighters to access a controlled water supply to attack fires. Fire hydrants are above-ground connections that provide access to a water supply. Their locations need to be planned and coordinated with the external works, including the provision of a pavement area that provides access and support to the fire brigade pumping appliance. In certain circumstances it may be possible for the design to make use of local reservoirs, lakes or rivers as the source of supply. Large premises, such as hospitals, universities or industrial plants may have privately owned water hydrants installed on site for use by firefighters in the event of a fire.

Hose reels provide a flexible means of directing water to precisely where it is required. Wet and dry risers are intended for use by firefighters to assist with fighting fires; they provide a means of of delivering considerable quantities of water to hose reels that are located throughout buildings. Wet risers comprise a system of valves and pipework which are kept permanently charged with water. Dry risers comprise a system of valves, pipework and pumps, but the pipework is usually empty and just enables firefighters to pump water to the upper floors of a building when required as illustrated in Figure 7.13.

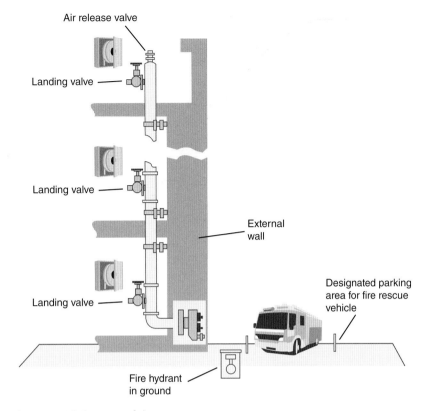

Figure 7.13 Provision of dry risers.

Sprinkler systems provide a fixed system to cover predefined areas which are automatically activated if a fire is detected and will douse an area with water. The system comprises a water supply system and pumps providing adequate pressure and flow rate to a piped distributed system terminating in sprinkler heads. The system incorporates valves, alarms and electrical supplies. Figure 7.14 indicates the main components.

Wastewater removal systems

This consists of drainage systems inside buildings to collect the wastewater and provide pipework distribution to connect to either a public or a private sewer, septic tank or cesspool.

The collection points include toilets, baths, sinks and basins, as well as drain points and floor gullies. Each system will be different, but salient points to consider are as follows:

■ Backpressure backflow is backflow caused by a downstream pressure being greater than the upstream. Systems are trapped to exclude smells and foul air. Traps are devices containing a water seal of about 2–3 cm, to prevent gases escaping into sanitary fittings: wash basins, water closets, sinks, baths, showers, etc.

■ In most cases the wastewater will fall under gravity, and sufficient fall should be allowed in the pipe design. However, if the gradient is too steep, self-siphonage may occur, where the contents of the trap are sucked out into the wastepipe because the water flows away too quickly, thus emptying the trap.

■ It is good practice to provide a vent for foul water drains. Any smells or pressure may be relieved at the vent.

Surface water may be discharged into public sewers alongside the wastewater. Alternatively, it may be discharged into a soakaway (an underground pit, filled with gravel, within the boundary of a building) or some other adequate infiltration system, or to a watercourse such as a river or lake.

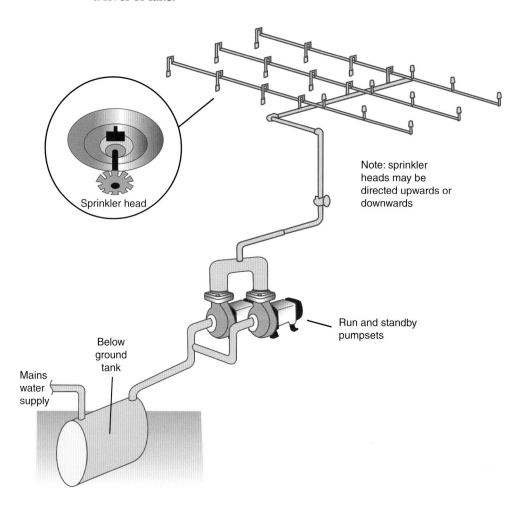

Figure 7.14 Components of a sprinkler system.

7.7   Gas systems

A gas distribution system distributes gas to its final points of use. There are a number of different gas systems that may be required, such as natural gas for space heating and cooling, water heating and cooking, and medical or laboratory gases, as well as gas powered generators used as backup power supplies. Industrial uses include waste treatment and incineration.

The term 'gas' is generally used to refer to 'natural gas' which can be provided from a public utility supplies national grid or in bottles. An alternative is liquefied petroleum gas (LPG) which is delivered from tanks and stored on site in special storage reservoirs (gas holders). These have particular requirements with respect to siting of the tanks, ventilation and conditions around the tanks, provision of access to delivery vehicles and security. Natural gas and LPG have different properties, meaning that LPG cannot be substituted for natural gas without being treated and mixed with air.

As illustrated in Figure 7.15, a gas distribution system has governors which control the gas pressure and flow rate pumps/compressors, a manifold to allow for gas to branch through several pipes, pressure regulators and shut-off valves, warning and alarm indicators and final outlets.

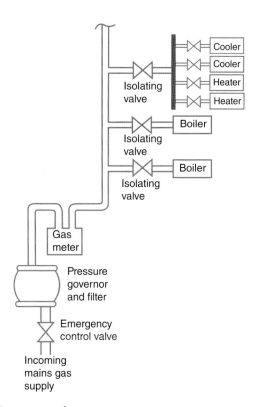

Figure 7.15 Components of a gas system.

Gas pipes leaking slowly into a cavity wall or other enclosed space is very dangerous. The gas leak may not be detected for some time, and an explosion can occur if the gas quantity builds up and a source of ignition is present. It is best practice for gas pipes to run outside as much as possible. If gas pipes have to run in an enclosed space a gas leak detection system with sensors, alarm and automatic gas shut-off valve should be provided – or if it is an enclosed vertical shaft an extract fan should be used so that the shaft is continuously fully vented.

LNG/LPG

LNG and LPG are sometimes used interchangeably. They are quite different things. LNG is liquefied natural gas, which describes the form in which natural gas is transported from its source to the destination where it is converted back into a gaseous state. LPG is a gaseous alternative to natural gas.

Medical and laboratory gas systems supply things such as piped oxygen, nitrous oxide, nitrogen, carbon dioxide and medical air to the points of use. These may either be standalone systems sourced from local cylinders or part of a piped system connected to central storage area. The same principles of regulating the pressure, distribution and emergency shut-off valves applies.

7.8   Electrical distribution

An electrical distribution system comprises a source or sources of power and a transmission system to the points of use. It will also form part of the building's earthing system.

Source of supply

The sources of electrical power to a building may comprise a single or a combination of public utility services suppliers, tie-ins to campus/site-wide networks, in-situ supplies, such as a diesel generator, or sustainable sources, including recycled 'waste' energy. Some electrical loads may require secondary or backup power supplies, which are used if the primary power supply fails. The voltage from the source may need to be transformed to the correct voltage for distribution in the building.

Transmission system

The transmission system comprises the fixed installation of cables, cable containment including their support and fixings, and distribution equipment which carries the electricity from the source to its final point.

Electricity is carried by electrical conductors. Direct contact by people or animals will give an electric shock, contact with other conducting materials will cause the electricity to leak to another material which, if touched, will also give an electric shock. Therefore the electrical conductors are either protected with electrical insulation or in certain circumstances, for example electrical pylons, it may be feasible to install the uninsulated cables as they are deemed to be safely out of reach. This insulation does not provide protection against mechanical damage, chemical and electrochemical attack, damage by insects, animals and plants, so mechanical protection is provided by additional sleeving in the form of a metallic armour sheath or toughened plastic over the electric insulation or by installing the cables inside a containment system of trunking or conduit. In addition, fire protection may be provided to ensure that, in a fire, the cables do not emit toxic fumes into the environment. Figure 7.16 illustrates various cases.

The cables are either fixed directly to the building fabric or indirectly by the containment system.

Cables are sized according to the electrical load they are serving to ensure that sufficient current can be drawn and the voltage (which falls due to volt drop) is adequate. They are designed to carry a specific maximum current for the particular conditions under which they are installed: ambient temperature, grouping with other cables and method of fixing. If the maximum current is exceeded the cable could overheat and becomes a potential fire hazard. Thus, all cables are protected by a

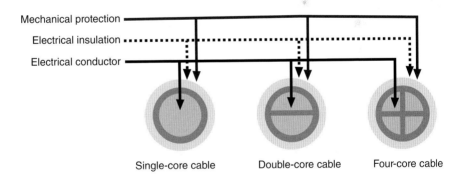

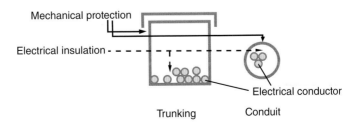

Figure 7.16 Examples of electrical cable and containment.

circuit protective device (CPD) which will automatically disconnect that cable from the distribution system if too much current is being drawn. These CPDs may also be used for manual switching for maintenance and testing.

The CPDs are generally grouped together in a common physical enclosure, along with meters and control devices to form electrical switchgear. These may be single-phase or three-phase depending upon their application. In all but the very smallest buildings, there will be many ways of subdividing and grouping electrical loads together. The final arrangement will depend on metering requirements, availability of space to accommodate equipment and cable routes, and minimising inconvenience in the event of an electrical fault. They will also be determined by the relationship between the supply sources and the final loads, for example if a load requires a backup supply. Figure 7.17 illustrates three ways of supplying the same final loads.

The cables will either terminate directly into the equipment they serve (e.g. luminaires) or into devices that provide an interface with the electrical loads they serve.

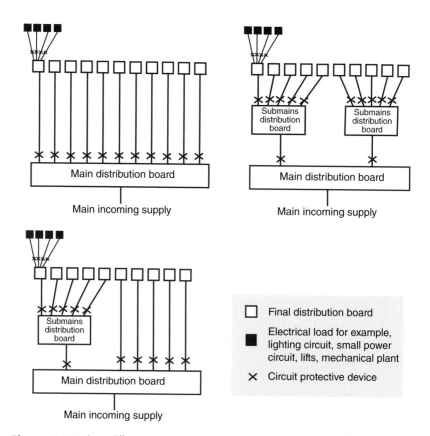

Figure 7.17 Three different arrangements for serving the same electrical loads.

This constitutes the fixed wiring system. In the case of socket outlets, portable appliances are connected into the fixed wiring system via plugs.

Earthing and bonding system

Should any fault develop in an electrical system the electricity will always head for earth, taking the easiest route: the path of least resistance. The terms earthing and bonding are often used interchangeably, but they are two quite different things.

The earthing system sits alongside the standard wiring system. The earth conductor, in electrical terms, carries no current but provides a direct path to a proper earth. Bonding is concerned with connecting items that are not part of the electrical system and might accidentally become live, for example raised metal floors, metal ceiling girds, water and gas pipes and large metal items such as a metal kitchen sink. Figure 7.18 illustrates this.

In additional, there is functional earthing, which may be provided as part of the proper functioning of electrical equipment. This would usually be a separate system to the building earthing and bonding system.

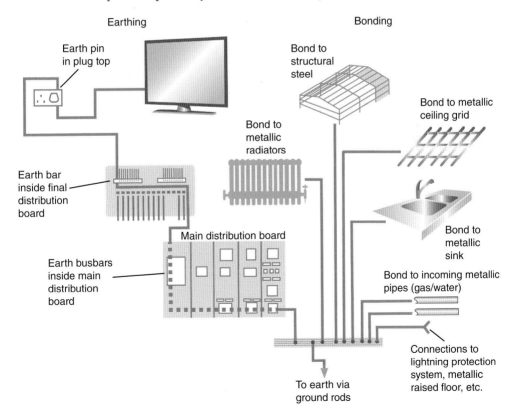

Figure 7.18 Basics of earthing and bonding systems.

Electrical supplies for mechanical, public health and other equipment

Some mechanical and public equipment will require electrical power supplies. Some will be motor-type loads, such as fans and pumps, while others will be resistive loads, such as air or water heaters, heating or cooling coils. In small buildings these may be fed from a local distribution board, but in larger buildings it is usual to cluster the power supply feeds together into motor control centre panels (MCCPs). MCCPs are like electrical switchgear in so far as they provide a convenient way of controlling a number of final loads from a common point. In addition, the panels may accommodate motors, starters and drive speed units.

Other equipment specified by other members of the design team may also require electrical power supplies and it is important for building services engineers to be aware of these: window cleaning equipment, roller shutter doors and automatic sliding or revolving doors and so on. On a smaller scale, electrical power supplies may be required for turning on or off water supplies to taps or for flushing toilets and urinals.

7.9   Artificial lighting

The artificial lighting system provides illuminance where and when required. This comprises luminaires and lamps, controls and the associated cabling and, where required, provision for alternative power supplies for use if the mains power supply fails. The artificial lighting scheme should be integrated with the daylighting design such that daylight is used when feasible, thus reducing energy consumption and possibly providing a more natural visual environment.

Relationship between lighting and space heating/cooling loads

Changes in lighting load density affect not only energy use for electric lighting but also energy requirements for space heating and cooling. In general, a reduction in electric energy use would tend to increase space heating during the winter months and lower the cooling requirement in the summer. The implications for total building energy use, however, would vary, depending on the building and building services designs, its operation and the prevailing climates.

The luminaires and lamps are selected on the basis of calculations to fulfil the design criteria. These take account of the room surface finishes, cleaning regimes and lamp replacement strategy. They will need to integrate with the building fabric and coordinate with the interior design requirements. Energy efficiency considerations are also part of the lighting system design.

Lighting control systems manage the outputs from luminaires to assist with creating the right visual environment and managing energy consumption. In the simplest form, full manual switching and dimming controls can be used. If the occupants are aware they can ensure that the manual system achieves the desired ambiance and avoid energy wastage by switching/dimming the system as required. Otherwise, more sophisticated systems with automatic controls and intelligent controls may be beneficial.

Automatic controls are either based on occupancy or illuminance levels in their area of coverage.

■ Occupancy sensors (including passive infrared, ultrasonic and dual technology sensors) serve three basic functions: to automatically turn luminaires on when a room is occupied, to maintain them in the on state, then to turn them off after a predetermined time or when the room is vacated.
■ Illuminance-level sensors responds to the ambient level and automatically turn luminaires on or off, or dim them to some preset level. Daylight dimming can maintain the desired light level while providing a smooth, barely noticeable transition to or from electric lighting as daylight increases or decreases.

These devices may be 'standalone' or they may be integrated into an intelligent system which can allow for monitoring and recording information, programming and remote control of operation of luminaires.

The cabling and its containment for both power for the lamps and the controls are sized accordingly and the routes planned.

With reference to Figure 7.19, emergency lighting systems are provided primarily to allow for illuminating a safe route out of a building if the mains power fails (escape lighting) or to allow occupants and tasks to continue operating if the mains power fails (standby lighting).

Once the areas of coverage are determined the design needs to consider:

■ whether the same or separate luminaires will be used
■ whether a central, local or individual battery units will be used
■ the method of testing.

Emergency escape lighting and fire alarms

Don't assume that emergency escape lighting is always about fire evacuation. The two are not necessarily coincident. Emergency escape lighting is designed to operate when the mains power fails, either to all or part of a building, in order (especially during night time) the occupants can safely leave the building – that's all.

A fire may occur which may or may not affect the electrical mains power supply. Of course, in a worst-case scenario, they may both occur together.

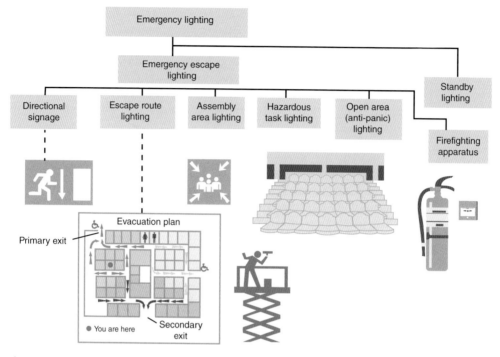

Figure 7.19 Components of emergency lighting systems.

External lighting

There are broadly four functional types of external lighting to be considered:

- sports lighting, specifically to allow the participants to perform their sport
- amenity lighting to allow users to move around an area
- landscape lighting to provide sufficient illumination to allow the plants and vegetation to grow
- architectural lighting to either illuminate architectural features or to be the feature itself.

7.10   Controls

Controls systems have a wide range of levels of complexity, the simplest being an on/off light switch, or time clock to control a piece of equipment. Thereafter controls systems use increasing levels of sophistication up to assessing the current and historic performance of a building/facility as a whole, and its significant energy-consuming systems and components. This allows decisions to be made to switch (on/off), modulate or adjust equipment to maintain some environmental condition and also to ensure that systems operate safely, efficiently and conveniently. More intelligent systems may incorporate self-adaptive control algorithms that automatically adjust set-points to suit the usage and load level.

Systems such as heating, ventilation, air-conditioning, lighting and access control use communications protocols which enable them to exchange information with other devices, but they do not always use the same protocols and so may not be able to communicate without further integration. An intelligent building management system is a type of open-platform software that brings these together into a single, integrated database which will allow different systems to work together.

A controls system comprises sensors, wiring and data acquisition devices, controllers and the means to calculate, display, report and archive the information. The hardware is supported by software, including protocols which determine how data is transmitted.

The sensors sense or measure a condition and actuators control the devices. These are linked together by a controller which takes the signal from the sensor and tells the actuator what to do. Alternatively the sensors could be providing information for monitoring purposes.

The controls system specification needs to define the intent of the systems. At the early stages this is usually done by means of a narrative describing the levels of control. This will cover the extent of end-user control requirements, building operators control functions, energy management issues, environmental conditions and control sequence logic. As the design develops, points schedules are required to define all the sensors for monitoring predefined parameters and the associated controlling devices for implementing the appropriate actions.

Protocols

Protocols define the set of rules by which devices communicate with each other. They may be closed or open. Closed or proprietary protocols are unique to a particular supplier's system. Any future changes and upgrades have to be done through them.

Open protocols are intended to be widely available and not tied to a particular supplier. Widely used open protocols in building controls are:

- BACnet, which is in widespread use throughout the HVAC industry and is an ISO standard protocol
- Modbus, which was first developed in the 1970s and is an openly published and royalty-free means of establishing 'master–slave' communication between intelligent devices.

It was designed mainly for industrial applications but is also used in infrastructure, transportation and energy
■ LonWorks, a networking platform created for control applications in 1999, Manufacturers in a variety of industries including building, home, street lighting, transportation, utilities and industrial automation have adopted the platform.

This is where an intelligent building management system (BMS) comes in. Information from controllers using BACnet, Modbus or LonWorks is directed via a router or 'gateway', which converts it into an open-platform ethernet or internet protocol (IP). Information from all systems is shared over a local area network, linked to the intelligent BMS itself.

The basic requirements for an automatic control system are:

■ facilities to monitor and measure the characteristics of the incoming utility services
■ facilities to start, set back and stop the plant
■ facilities to control the volumetric airflow or fluid flow rates
■ facilities to control the system or room pressure
■ temperature measurement, control and indication
■ humidity control and indication
■ devices to monitor and indicate the plant's operating state
■ alarms to indicate plant failure, low airflow and filter state.

The controls system may be connected to the graphic interface, which allows the user to navigate and interrogate the system using a web browser. Through this interface they can view analysis, graphics, trends and reports, and perform functions such as updating lighting schedules and managing alarms.

7.11   Lightning protection system

Lightning protection systems prevent a lightning strike from directly or indirectly having a detrimental effect on the operation of a building. A risk assessment determines whether a system is required and, if so, the level of protection required.

The system comprises an air termination network, which deliberately attracts lightning to itself, in order to direct the associated electrical current away from the building via a network of conductors. These are connected to earth where the currents are dispersed.

7.12 Fire detection and alarm system

Fire detection and alarm systems monitor for potential fire hazards and react accordingly. The system comprises inputs, processing capability and outputs.

Building services engineers' input to the fire detection and alarm system should respond to an overall strategy for the building. The overall fire engineering strategy comprises a risk-based evaluation for a project which will take into account issues such as building fabric and materials, building configuration and provision of adequate escape routes, operation of building services systems in a fire, including firefighters' control over ventilation systems in fire conditions. It should take into account the requirements of the buildings insurers who will be interested in protecting their value in the building and the contents.

Figure 7.20 illustrates some of the potential inputs and outputs.

For the fire engineering strategy, building services engineers will need to understand:

- fire compartmentation
- level of automatic detector coverage
- requirements for dealing with the hard of hearing
- fire service access location – in order to locate the main fire alarm panel
- location of repeater fire alarm panels

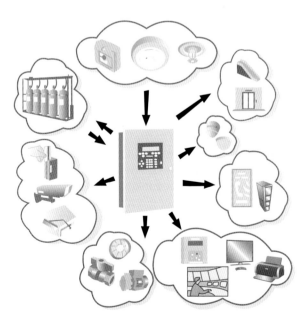

Figure 7.20 Inputs and outputs to a fire detection and alarm system.

- evacuation strategy – to develop the cause and effect matrix which details the sequencing of alarms and actions of other outputs for any particular inputs
- backup power supply arrangements – to provide the appropriate battery
- interfaces with specific areas requiring supplementary fire suppression systems, which may comprise a combination of dry chemicals and/or wet agents to suppress equipment fires. Common means of detection are through heat sensors, wiring or manual detection (depending on system selection)
- requirements for firefighters' operating panel – these allow the firefighters to control aspects of the building services systems, for example to shutdown ventilation plant, to operate fans associated with smoke extract, and switches to isolate high voltage luminous signs or to cut off electrical power to certain equipment.

7.13   Smoke and fire control systems

Smoke control systems ensure that if there is a fire, smoke is contained and removed, allowing safe evacuation by clearing smoke from the designated escape routes, and firefighting by maintaining smoke-free areas for firefighters, and minimising damage to the building and its contents by controlling the build-up of heat and smoke.

It can include smoke and fire dampers, ductwork, smoke curtains, fans which may be integrated into the design of the ventilation system, and interfaces with other devices which may need to operate such as doors and roller shutters. The electrical power supplies to devices associated with the smoke and fire control systems will need to be provided with a backup power supply for use if the mains power supply, and cabling needs to be fire-rated and routed to reduce risk.

Building services engineers need to understand the building compartmentation and location of the escape routes. A fire compartment is a building space or area that is separated from other spaces by walls and floors that offer some level of fire resistance. If a fire starts in a building, small openings in fire-rated walls, floors and ceilings for wires, pipes, ductwork and other building components can become open channels for fire and toxic smoke, endangering both the structure and its occupants. If ductwork penetrates a compartment of a building, it must be designed and installed so as to contain the spread of fire. Figure 7.21 illustrates a simple example of smoke dampers, which would operate to stop smoke spreading.

This principle of fire stopping openings where ductwork penetrates fire-rated walls and floors also applies to cables and cable carrying systems (trunking, trays and conduits). These are often fire-stopped by sealing with intumescent, endothermic or ablative materials.

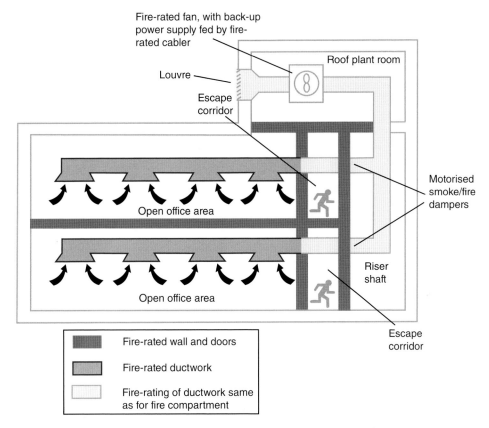

Figure 7.21 Ductwork and fire compartments.

Intumescent materials, which expand when heated, can be used to prevent the spreading of fire. Endothermic materials release chemically bound molecules of water under heat, as temperature increases, and water is driven out of the material in the form of steam, which cools the surrounding surfaces. Ablative materials are typically silicone-based and help resist fire by slowly eroding and forming a char when exposed to high temperature conditions.

When building forms and layouts do not fall into standard categorises, as covered by the regulations, fire engineering solution may be required, such as in atrium spaces shown in Figure 7.22.

Atriums are a focal point of a building, combining escape routes, waiting areas, main entrance and perhaps retail. A possible fire engineering solution would be a fire-rated, sterile box, with no balconies and no reception desk in the area. However, it may be possible for building services engineers to 'prove' that a more relaxed scheme works by means of computational fluid dynamics (CFD) analysis to predict the movement of smoke from possible fires in complex enclosed spaces. This is especially important as, unlike with other building systems, it is virtually impossible

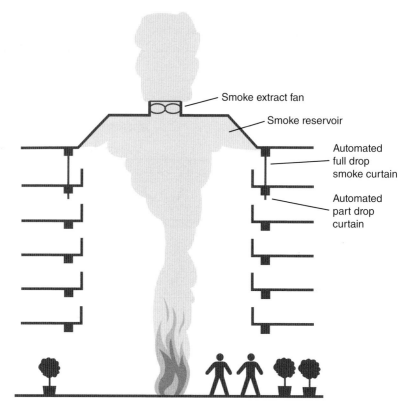

Figure 7.22 Smoke control in an atrium – a possible solution.

to test an atrium smoke-control system in real conditions. This is primarily because real conditions involve large fires that can damage an atrium. Design modelling can also include wind, again for which systems are nearly impossible to test.

Sprinklers

Fire and rescue services are mostly concerned with ensuring that, during the fire, the building can either be safely evacuated or people moved internally to a safe place and that the fire personnel can safely enter the building and control the fire. Once the fire is extinguished the building is safe. They are not concerned with the property within the building or the long-term prospects for the structure.

So the local enforcing authority may not require a sprinkler system to be installed, but the building insurers may want it. Thus is it important for building services engineers to liaise with the building's insurers.

7.14   Security systems

Security systems provide surveillance for people, the building and the contents within, against damage and loss. The overall security strategy for a project takes into account issues such as access and movement of people, vehicles and goods, building structure, site boundaries, physical protection, types of activities, management and maintenance and use of manned and animal-supported surveillance. This will be based on a risk assessment of potential threats and vulnerabilities. It should also take into account the requirements of the building's insurers with respect to the protection of the building and its contents.

Building services engineers' input to the security system should respond to an overall security engineering strategy for the building. This may include security lighting, access control system, closed circuit television, alarm system and a patrol system.

Security lighting

Security lighting allows those entering a space to be clearly identified to determine whether they are a 'friend' or a potential intruder, to deter intruders or to increase the feeling of safety, for example when walking across a car park. Security lighting needs to be designed and coordinated with interior and exterior lighting systems. The security strategy needs to identify areas and their requirements for illumination and controls for security reasons.

Access control system

This allows or denies individuals and vehicles entry to defined physical areas in accordance with predefined criteria. The simplest form would be an intercom and door-release, with the decision as to whether to allow entry being made locally or by providing PIN codes to selected people. More complex systems identify the requester by means of card-readers, chip-readers and electronic locks that read information encoded on the cards, discs or keys carried by requesters. This may use insertion- or swipe-readers that interpret magnetic stripe cards, or proximity-readers that do not require physical contact with the cards they read. More sophisticated systems incorporate biometric devices based on fingerprints, voice-prints or retinal patterns.

Any system that requires automatic door opening needs to consider the strikes, contacts and releases that operate doors. The complete system may include the software for monitoring and recording movements, managing the distribution and encoding of cards and the processing of transactions.

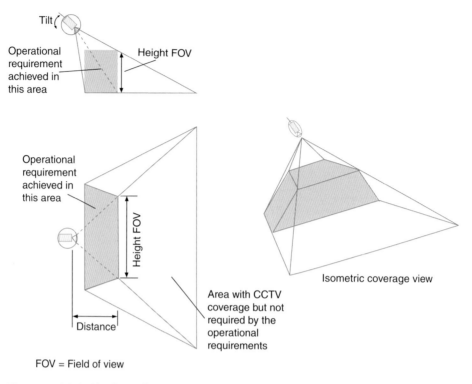

Isometric coverage view

Area with CCTV coverage but not required by the operational requirements

FOV = Field of view

Figure 7.23 Fields of view for CCTV cameras.

Closed circuit television

This can be used to visually survey, from another location, defined areas in order to enhance surveillance of a defined area of coverage. The system requires cameras, processing capability and a means of viewing the images in real time, or to be stored for later viewing or both.

The security strategy needs to define the fields of view, as illustrated in Figure 7.23, the quality of the image, whether night vision is required and whether the surveillance is overt or covert. This will inform the lens type and whether fixed or pan tilt zoom (PTZ) lens is required.

The security strategy also needs to identify which images need to be viewable in real time and which need to be recorded, and also the quality (in frames/second) and duration (how many days/hours/months) of recorded material.

Alarms

These are used to monitor for intruders in a defined area, or disturbances to the building fabric. They consist of sensors and a means of indicating an alarm condition. Sensors can detect patterns in motion, presence, noise and vibration. Response to a sensor operating may be operation of sounders and/or lights, or relaying information to a local or remote control centre or dialling out to a phone or pager. In a very simple system the sensors may be directly connected to these devices, while increasing levels of complexity require more processing capability. In addition there may be alarms for occupants that operate in an emergency.

The security strategy needs to identify the areas and items to be monitored and the actions required.

Patrol stations

These allow for recording the locations of personnel against predetermined criteria. If there are deviations from the criteria an alarm will be initiated. This requires a means of detection and a data logger. The security strategy needs to identify locations of the checkpoints.

Security systems may be interfaced with each other and with other systems, such as access control system (to disable doors in the event of an intrusion), lighting system (to switch on luminaires in response to a detection), electrical system (for power supplies to devices) and motion detectors that detect movement on video signals transmitted from closed-circuit TV (CCTV) cameras.

The routing and protection of cabling needs to be considered to minimise the potential for accidental or premeditated damage, which could affect the integrity of the security strategy for a building.

7.15 Structured wiring system

A structured wiring system is a complete system which provides the infrastructure to support the transmission of data associated with ICT systems. It is often referred to as a 'passive' system. This infrastructure is used by a wide range of 'active' devices to distribute data.

Structured wiring systems comprise cables and patch panels (which act as junction boxes and distribution points), culminating in outlets into which appliances can be plugged. The system is arranged to support the required transmission protocols of the active equipment which uses it. Figure 7.24 illustrates the main components.

There is more to cabling system performance than just buying good quality cable and connecting hardware. In order to ensure full bandwidth potential, a structured cabling system needs to be properly designed, installed and tested. As such most structured wiring systems are warrantied as complete end-to-end systems. This covers the design, installation, testing and commissioning from a single source.

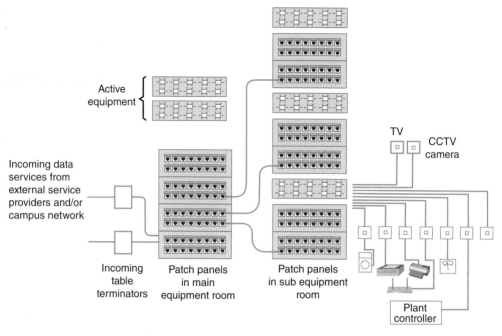

Active
equipment

Incoming data
services from
external service
providers and/or
campus network

TV

CCTV
camera

Incoming
table
terminators

Patch panels
in main
equipment room

Patch panels
in sub equipment
room

Plant
controller

Figure 7.24 Components of a structured wiring system.

7.16 Broadcast communications technology systems

Broadcast communications technology systems receive audio and video
one-way transmissions via a dedicated receiver. The receiver decodes
the signals into a suitable format to transfer the signals to final outlets
into which portable devices can be plugged. Figure 7.25 illustrates various likely scenarios.

Building services engineers need to consider the locations for the
receiving equipment and modems, cable distribution routes and
final outlets. The receiving equipment and modems may require
electrical power supplies.

7.17 Mobile telephony systems

Most buildings need to be able to receive mobile phone signals. Different
service providers will have different coverage and signal strength.
Although they will provide a 'coverage checker' based on a location,
this may not represent the coverage in a new building as the building
acts as a shield to mobile signals, and areas deep in a building – say in
underground car parks – may have particularly poor reception. Also, if
the new building increases the local usage, the existing system may

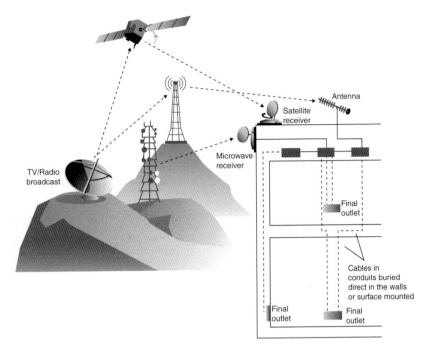

Figure 7.25 Principles of broadcast TV components.

become overloaded with the new demand. Therefore the mobile phone service providers should be consulted for advice.

The mobile services providers may be able to strengthen the existing signal from the nearest cell site with no impact on the building. However, it may be necessary to install a new mast on the building or install internal local boosters. Alternatively, for areas such as tunnels a leaky feeder system may be considered. This consists of a coaxial cable which emits and receives radio waves, effectively functioning as an extended antenna. All of these require coordination by building services engineers in terms of space allocation and any electrical power supplies. A typical set-up is shown in Figure 7.26.

7.18 Audio, visual, audiovisual and information systems

Audio, visual, audiovisual (AV) and information systems use ICT technologies to support the use of the systems. Examples are

Audio

■ public address systems
■ music systems
■ simultaneous translating systems

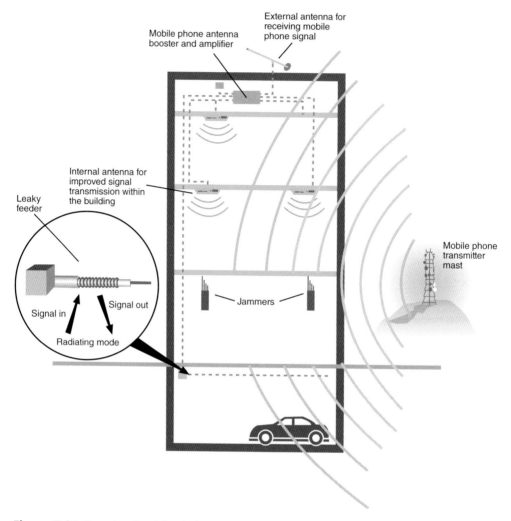

Figure 7.26 Principles of mobile telephony systems.

Visual

- projectors
- LCD or plasma screens displaying broadcast or in-house television services
- rolling displays
- voting systems
- centralised clock

Audiovisual

- video-conferencing facilities

Information systems

■ real-time locating
■ RFID asset tagging

These systems may be interfaced with

■ electrical distribution systems for connection points
■ fire detection and alarm systems.

In addition there are multimedia systems, which use a combination of different content forms. Building services engineers need to understand the space and any adjacency requirements for each of the systems.

7.19   Facilities for the disabled

Facilities for the disabled are provided to support building users with disabilities. Building services engineers' input to these systems should respond to an overall strategy for the building for people with disabilities which will take into account issues such as access and movement. However, disabilities are not just confined to wheelchair users, meaning that it is not all about providing ramps for wheelchair access but includes facilities for those have other physical impairments: sensory, such as sight or hearing, or mental impairments including learning disabilities. Building services engineers will need to consider:

■ lighting, which is essential for everyone for visibility and safety but more especially for people with disabilities, who need extra assistance with navigation and with reading way-finding signage. These need good illuminance levels, uniformity and colour-contrast
■ audio induction loop systems in areas where background noise may be a problem. These may be transient areas, such as the reception area and ticket counter, or extended time use areas, such as auditoriums and meeting rooms
■ vertical transportation to upper floors with controls suitable for wheelchair users and people with sensory impairments
■ provision of appropriate toilet and washing facilities, including the sanitaryware and taps and their controls
■ light switches and other controls within reach of wheelchair users and those with sensory impairments
■ consideration of alerting the need for emergency evacuation and the procedures involved with the evacuation. This may be aided by fire alarms fitted with flashing lights, providing a vibrating pager or

pillow (for sleeping areas) to alert hearing-impaired people and using pink noise for emergency escape routes
■ telephones in disabled refuge areas, to provide two-way communication between the area and the designated control area.

Provision for people with disabilities should be considered as an integral element of the design.

7.20   Vertical transportation

Vertical transportation systems support the movement of people and goods between different levels of a building. Building services engineers' input to these systems should respond to an overall strategy for the building to give the required quality and quantity of the vertical transportation service provided, in terms of waiting times are and what percentage of the building population can be transported by lifts or escalators in a given time period. Consideration is required in respect of the lift shaft space requirements and any associated plant rooms. Building services engineers will need to consider:

■ electrical power supply requirements to suit the requirements of suppliers, including power for the motor drives and control panels
■ lighting and emergency lighting for the lift cars, shaft and plant rooms
■ interfaces with the fire detection and alarm system. This includes provision of automatic fire detectors and alarms, and inputs/outputs
■ any interfaces with the buildings access control system
■ any interfaces with the buildings CCTV system
■ any interfaces with the buildings control systems; for example for the purpose of monitoring and remote control of the lifts or escalators.

Summary

The building services engineering systems individually and collectively contribute to creating the necessary conditions to satisfy the occupants and processes in a building. The design process embraces technical principles, recognition of legislation, standards and good practice, but these include varying degrees of prescriptive elements. This allows designers to make use of creative input and to choose components from the plethora offered by suppliers. This means that the final solutions by different building services engineers may be different, but still correct for achieving compliance with design criteria and coordination with other services and the building fabric.

8 Off-site manufacturing

Off-site manufacturing covers the manufacturing of sections of a building, then transporting them to site for final positioning and connections. Advantages are seen to be better health and safety, higher speed of construction, lower costs, higher potential for zero defects and reducing the need for skilled labour on site.

Prefabrication refers to the assembly of components. Preassembly refers to the assembly of several components to form a complex unit. Modular construction refers to the assembly of multiple discipline components, such as a boiler room in a weatherproof enclosure or a bathroom pod.

The building services could either be integrated into components or assemblies, mainly intended for the likes of the building structure, fabric or finishes, for example:

■   cast-in conduit
■   self-contained rooms, e.g. bathroom pods
■   curtain walling
■   penetrations through steelwork
■   earthing in piles

or they could be standalone building services assemblies or components, for example:

■   a combination of services pipework, ductwork, cable trays/trunking modules used in horizontal or vertical distribution
■   pumpsets complete with pipework, meters and valves
■   packaged plant rooms.

Building Services Design Management, First Edition. Jackie Portman.
© 2014 John Wiley & Sons, Ltd. Published 2014 by John Wiley & Sons, Ltd.

To achieve this, considerations are required during the design stage with respect to:

- competency and experience of designers in off-site manufacturing
- extent of standardisation/repetition/symmetry of building
- building tolerances affecting the integrity of the interconnections and interfaces
- transportation constraints: be aware of the width, height, length and weight limits for vehicles and plan the route, to avoid constraints due to low bridges, and width and weight restrictions
- setting in position needs to recognise any temporary storage on site and cranage to final positions. This needs to coordinate with the programming/construction sequence
- deconstruction, dismantling and recycling after use.

Summary

Off-site manufacturing offers an alternative way of progressing the construction works, with benefits associated with health and safety and possible cost and programme benefits.

Part Three   The design management process

Design relies upon a combination of scientific principles, established standards and practices combined with varying degrees of aesthetic and human factors applied in a well thought out manner.

The design management process provides a framework to understand requirements which are used to develop and deliver design information for use by the construction cohort. The design for any particular aspect of building services systems involves many organisations and is characterised by its iterative nature, so a design management process, which is part of the overall project management structure, should be followed irrespective of the type of commission/projects.

Project management is about guiding resources throughout the life cycle of a project, using appropriate tools and processes to achieve predetermined objectives relating to scope, cost and quality within a finite timeframe. It requires an understanding of the construction and design process as well as knowledge of areas of business management: risk management, information management, programme management, value management, cost management, programming, quality management and design management.

9 Design execution

The design execution process starts with the initial recognition of a need and ends with the maintenance of the completed building. The earlier stages are more involved with architectural concepts but building services issues should be considered: establishing external and internal design criteria, determining sizing and capacity requirements, considering options and determining those most suitable, all within constraints that apply, financial or otherwise, and always in collaboration with other members of the team. As the design proceeds, the emphasis is more on producing the deliverables needed for defining the construction works.

Building Services Design Management, First Edition. Jackie Portman.
© 2014 John Wiley & Sons, Ltd. Published 2014 by John Wiley & Sons, Ltd.

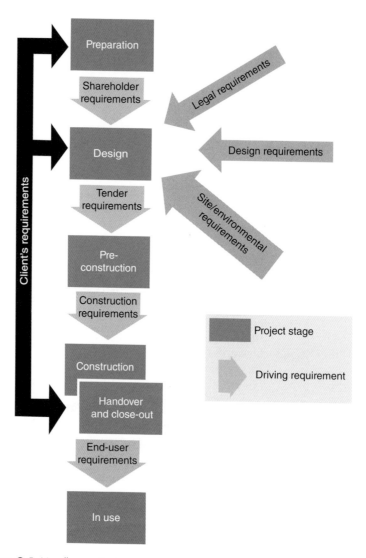

Figure 9.1 Headline project stages.

Design execution usually makes use of models, processes, procedures and tools and it aims to prescribe how things must/should/could be done and to provide tools to help monitor progress. Models of the design process can be descriptive (describing a succession of activities or thought processes), prescriptive (aiming to prescribe an improved approach to the process of design by structuring design behaviour) or consensual, using a systems engineering approach to combine descriptive and prescriptive models. The use of particular model would be defined in the building services engineers' contract. Standard model templates are produced by various bodies from different disciplines, industries and interest groups. This book deliberately avoids following any specific model: an important part of the building services design manager's roles is to understand the process including the inputs, interfaces and deliverables required at each stage. Instead broad headings described in Figure 9.1 are used.

The exact requirements for each stage will vary according to the details of a contract, thus it is vital for the contract to be reviewed at each stage.

Salient principles applicable to all disciplines include realising that:

■ project stages may have different names on different projects
■ not all project stages may be required for a project, or building services engineers may not (rightly or wrongly) be involved at a particular stage
■ deliverables and sign-off procedures will vary from project to project, so it is imperative that all parties are aware and in agreement.

While some aspects of building services design management are common to all the design disciplines, certain aspects need to be considered differently with respect to building services engineering.

9.1 Project stages

Preparation

At this stage, clients will have formally, or perhaps informally, undertaken a business case analysis to determine if a project is to be pursued. This identifies their requirements, the project objectives and the possible constraints on the project.

Initial client briefs, including any related feasibility studies, are usually prepared by architects. This will respond to the project objectives set out in the business case. They may consider possible sites and options on refurbishing or extending existing buildings, and preliminary thoughts on procurement routes. These help clients to decide whether to proceed with projects and to help with selection of procurement methods. The latter is a particularly important decision, as it will determine the way in which project resources, responsibilities and risks are apportioned between the clients, the design team and the construction team.

Ideally, building services engineers should be involved at this stage, particularly if there are significant issues relating to the building services engineering that may affect the viability of the project, for example the availability of sufficient utility services, in-situ utility services that need to be moved and environmental consequences of the project. They may make recommendations for work in later stages, such as intrusive site surveys. It may be prudent, at this stage, to consider any lessons learnt from previous similar projects. More realistically, building services engineers may not be appointed at the preparation stage, and will be in a catch-up position when they are. However, whatever the client's brief, at the outset it usually has to be developed further to provide a full design brief or output specification.

Clients know best

From the point of view of the building services systems, what clients really want is something that delivers appropriate conditions. Therefore when you're looking at the requirements from a client, they of course want the building services engineering design to deliver comfortable working and living conditions to enable the business to function efficiently. But it's highly unlikely that the client is going to say, 'What I actually want is Joe Bloggs' radiators and I particularly like the ones with the gold paint finish etc.'. They're not interested because, to them, the design is just a means to an end. And this is one of our problems, because although clients want the conditions to be right, building services engineers want to concentrate on the pipes, the ducts, the bits of equipment and so on. So in essence, the client isn't interested in the hardware that does all the work, he just wants the hardware that creates the space to be right.

Design

There will be several stages to the design phase, during which the level of detail considered and communicated in the deliverables will increase. Terms used include concept, scheme, detailed design and design development; sometimes, for smaller or fast-track projects, some design stages are omitted or for more complex projects additional stages are added. During this phase all the design disciplines will be developing their respective designs and will need to interface with each other.

The major task of conceptual design is to generate design concepts, evaluate them and select one or more best concepts for further development in the subsequent stages. At scheme design, solutions for systems are described and the disposition and approximate size of the plant room and distribution space allowances are determined. At detailed design stage, the particulars of the systems are developed, down to the component level and these are coordinated fully with the building fabric and structural solutions.

The exact requirements of each stage will vary, according to the individual contract and, in reality, building services engineers may continue to the next stage with matters outstanding on the understanding that they will be closed out; however, poor performance at one stage can rarely be compensated for at later design stages.

Design is a process of making choices. Selecting between the possible options at any stage for a given design, where all the potential solutions satisfy all the design criteria, becomes an issue of preference. This may occur in a design team – each member may have different preferences for addressing the same issue. It may be necessary to provide a ranking system to determine the best outcome for the project. Proper selection is important because poor selection can rarely be compensated for at later design stages and it can incur great expense in terms of redesign cost.

The importance of informal processes in design performance should not be underestimated. Knowledge, ideas and customer satisfaction, although difficult to measure, are often central to design performance. At the process level, informal discussions in the coffee lounge or at lunch can sometimes be more important than formal meetings.

The following are key activities that progress during the design stages:

Contributing to developing the client brief

The briefing process starts with a broad statement of intent during the preparation stage, and continues through the early design stages to a point prior to detailed design, when the consolidated brief should be agreed and frozen between the client and all the contributors to the project. An important objective of producing a consolidated brief is to allow the design team to design what the client requires and to avoid abortive work arising from late changes.

The level of information obtainable from clients will vary according to their experience, the form of contract and the skills of the building services engineers. Key information that needs to be covered is:

- the rationale for the project
- functional and operational requirements
- health, safety and environmental aspirations
- commercial policies associated with balancing capital and operational costs
- programme constraints, including any requirements for phased handovers.

Clients can express aspirations, which building services engineers need to interpret into design criteria and verify as being achievable.

Space planning for building services engineering equipment

This will start with plans for the main rooms, routes and risers to accommodate building services engineering equipment. At this stage, there will be limited information, if any, about the equipment that will be installed. These will be based on published rules of thumb and the building services engineers' experience. During the design phase these will become firmer and will be proved by producing drawings that show coordination with other services and the building fabric.

A spring day

A client brief for an indoor cafe might be as 'helpful' as 'I want the cafe to feel like a spring day'. This would lead the building services engineers to make personal judgements over what a spring day means; for example:

- temperature – records show a recorded temperature range of –6 °C to 23 °C
- ventilation rate – these could be still or up to 10 m/s on a breezy day
- illuminance levels – it could be over 50,000 lux
- acoustics – tweeting bird sounds might be expected, and perhaps traffic noises.

Of course, the building services engineer will have to ensure that the design complies with relevant statutory requirements, but even so, there is still fair amount of latitude for different solutions.

Building services engineers need to ensure that there are sufficient rooms, routes and risers for the equipment. This includes space for access, maintenance and future replacement and spare capacity for future developments. If too much space is provided, additional unnecessary costs will ensue, both due to the increased capital costs in building and because the extra space might eat into lettable space, reducing income. If too little space is provided, the equipment will not fit, or, if it can be squeezed in, its functionality or maintainability may be compromised.

Getting it right

The problem for the architect and the client is that all the ceiling and floor voids on a high-rise building add up … and so we again have to battle for distribution space. The difference between a 700 mm deep ceiling void and one which is only 500 mm deep may not sound very much. But when you've got 20, 30 or even 50 floors and you shave off 100 mm here and there, you're losing a potential lettable floor, particularly where planning permission limits the overall height of the building!

The client even reasons that by cutting down the plant room space, even by a little, then that's an extra few square metres that can be rented out.

So there is always a commercial balance required between needing adequate space for building services and building function.

Plant rooms

At the outset, the size of plant rooms is usually estimated on the basis of experience and using recognised 'rules of thumb' documents. As the design progresses, these can be further refined. Ideally, from a building services engineering design perspective plant rooms should be located at the centre of the zone that the equipment serves because this reduces the amount of horizontal and vertical distribution subsequently needed. In reality, this is modified by architectural objectives, structural solutions, adjacency requirements, connections to distribution systems and access requirements.

In most cases it is better to have clear open plant space: columns may obstruct operational space associated with the plant (e.g. to operate switches or to withdraw components – filters in air-handling units, circuit breakers in switchgear) or compromise access/escape routes – beams may obstruct cranage or cross-plant room service routes – and imperforable walls (those incapable of being perforated or bored through) will restrict how we can distribute the connecting incoming and outgoing distribution services (pipes, cables and ductwork).

The weight of the plant needs to be coordinated with the design of floor slabs. Floor slabs other than ground floors may not necessarily be designed to support heavy weights particularly items such as lifting gear attached to their underside. If the slab is required to bear fully loaded lifting devices, provision should be made in the structural design. The access and egress of the plant both during initial installation and during its lifetime (for maintenance and repair) and ultimate disposal needs to be planned for. During initial installation it may be reasonable to allow for the plant to be offloaded and put into place via a temporary opening in the building fabric. On disposal it may be possible to consider that the plant will be dismantled in-situ and removed in sections. In which case, the largest or most awkward section needs to be considered with respect to egress routes to a suitable external location.

Adjacencies to other related plant rooms need to be considered to minimise the lengths of pipes, cables and ductwork, otherwise if these links have significant length other building spaces may be compromised. There may be for practical reasons, if a particular plant equipment is connected to another (by pipe, cable or ductwork) it makes sense for them to be located in adjacent rooms. If plant equipment is connected to incoming utility services, they should be located as near as possible to their intake positions.

The heights of plant rooms need to be considered. The points where services (pipes, cables, ductwork) exit the room will be the most congested, which will limit the number of services that can exit to an adjacent riser without compromising access. This may be the determining factor for the height requirements – rather than the actual plant itself.

Space required for future plant must not be compromised. As well as floor space, the design needs to consider delivery, off-loading, putting into position and connecting up. This should ensure that access to existing plant is not then compromised. It this involves cranage, this should be planned for; it is not good enough to assume that it is possible to locate a temporary crane at the side of a building – there may not be sufficient space for the size of plant involved, or it may not be possible to get clearance for road closures.

Access arrangement needs to be considered. It may not be desirable for operations and maintenance personnel to enter plant rooms via the main building areas. Indeed, for access to plant rooms by utility services providers, it may not be allowable at all.

When designing a plant room, any other equipment that may be in that room needs to be coordinated with the layout. This may include space for items such as breathing apparatus, ear defenders, eye washers and fire extinguishers, and also, any signage, schematics and wall charts that are required to be wall mounted.

Much of the heavy plant equipment ends up in the base of a building. There are a number of practical reasons for this, but there are some exceptions, such as gravity operated water tanks and air-handling units which are often found at the top of a building.

Risers
Building services engineers should ensure that the distribution systems (pipes, cables and ductwork) are coordinated in their routeing throughout the building, and that the central plant room connects logically with the horizontal and vertical distribution routes.

Horizontal distribution may be exposed or concealed within floor or ceiling voids. Vertical distribution may be exposed or in dedicated riser shafts or concealed within the building fabric. Often the transition between horizontal and vertical distribution is the most congested area of the service core. The connection of horizontal ductwork or pipework with vertical risers should also be carefully considered. If the service core is enclosed on three sides by lift shaft and external walls, horizontal distribution from the service core will be extremely difficult and will provide little space for installation access and maintenance.

In false ceiling voids or riser spaces there are several distribution systems vying for restricted space, and they will need to be carefully coordinated with each other, the building structure and the components of the ceiling to ensure ease of installation and access for maintenance repair and removal of system components. This includes consideration of access panels in the ceiling – if the ceiling is not fully demountable.

The design of the distribution system routes needs to recognise any constraints imposed by the building's structural solution and fire compartmentation strategy. The structural solution may impose constraints on the size and location for distribution services to pass through. The fire engineering solution may dictate where special treatment (e.g. fire dampers) are required when services pass through fire compartments.

Space planning considerations for a heating system

Firstly, where do we place the equipment? For any system one of the simplest ways of thinking about this is by way of a liquid density gradient – in other words, the heavy stuff sinks to the bottom, and the light stuff rises to the top. There's no point in having lots of extra structure at the top of a building, just to locate heavy equipment that could be just as easily located at the basement.

Building services plant, as we will see, mostly consists of large boxes, usually made of metal, some being much heavier than others. In fact, some of the heaviest items that go into any building are the water storage tanks. Anyone who has ever had to move a fish tank or an aquarium will have a good idea of how much even a small quantity of water weighs. So wherever possible we put things that contain water – boilers, hot water storage, cold water storage, etc. – low down. Of course, we can't always do this with water because some systems require gravity distribution to work, and so water tanks often tend to be located at high level.

But boilers certainly tend to go into the basement – also for another very good reason. Today the most common fuel used for heating is gas. Now, if water leaks out of your system, it's a nuisance, it makes a mess of things but that's about it. But if gas leaks out of a system, it's potentially dangerous, so the less gas you have running round a building, the better, and the shorter the run between the incoming gas main and the boilers, the safer is the whole system.

And not just for gas: the same principle applies to any fuel – oil, coal, biomass woodchips etc. – the less you're trekking these round the building the better. So access for storage, delivery etc. goes to the basement.

As for the remainder of the space requirements, with most heating systems it's mainly a question of distribution, probably just pipes and emitters to get the heat out into the space. So it's not going to take a huge amount, but even with this you're probably using 6–10 per cent of the available building space. In reality, building services engineers will have a rough idea from previous projects as to the space required, to which they will add a margin, just to be on the safe side. But architects and others will think, 'They'll probably add a margin' and therefore they ask it to be cut down by 20 per cent!

Helpfully, you don't need a ceiling void but you do need plant room space, distribution space for the hot water pipes, such as a vertical riser, and a chimney or flue for the products of combustion.

Coordination with the building fabric and structure

Typically, the building services engineering designs need to fit into a frame whose design is predominately led by architects and structural engineers, with space allowances only for building services engineering systems. Traditional administration of building services engineering systems regards them as having confined scope, operating in isolation or tightly coupled and providing minimal support for overall

---

Space planning considerations for ventilation system

As with most building services the reasoning behind this is quite simple. If you think about what you want to do, you're going to bring air into your building and then you're going to blow it round your building. But you obviously want your air to be clean. And if you bring air in at say basement or ground floor level, you're liable to bring in lots of traffic fumes and pollution, which are actually quite difficult to deal with in terms of filtration. Whereas if you bring air in at a higher level, it's going to be much cleaner because many of the pollutants are quite heavy and stay at low level.

So if bring your bring air in at a high level, it's going to be cleaner. Notice we don't talk about being clean, but cleaner. So traditionally air intakes tend to be at high level.

Now if you're going to be cooling this air you'll need refrigerant – a sort of chemical liquid – which, if it leaks as you run it around a building, could cause potential problems. So any cooling equipment tends to stay close to the air intake so that we can cool the air down as quickly as possible. So in terms of vertical distribution, it means that cooling plant tends to be located at high level and boiler plant at low.

So just thinking about basic location of equipment within a building, it means that you require at least two plant rooms, one at high level plus one at low level, together with the necessary vertical and horizontal distribution space that they require.

---

coordination and holistic management hindering the provisioning of advanced services, and there are tensions between the size of these building services spaces, usable floor space and ceiling height.

Spare capacity and design margins

The terms spare capacity and design margins are often, incorrectly, used interchangeably, but they not the same thing. Spare capacity might be allowed for the following reasons:

- Future physical growth in the building's footprint, such as additional storeys or extensions as illustrated in Figure 9.2. It is important that the building is designed from the outset to accommodate such changes.
- Coping with changes in use that demanded more capacity – the pattern of use within a building over time is almost certain to change. Changes may be driven by many reasons, for example, from competitive pressures and customer requirements, new management approaches, new technologies, changing fashions or changes in regulations. These have direct consequences for the way workplaces are organised, for example encouraging changes in work styles

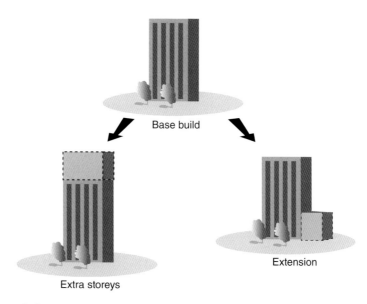

Figure 9.2 Spare capacity.

between individual, shared group or team-based workspaces, which result in changes in occupancy density, occupancy hours and building use such that the new use requires changes to the existing utility services.

■ A change in facilities services, for example, the provision of additional catering or shower facilities and, in the longer term, adapting office buildings for changes in building function, for example, refurbishment for residential use.

■ Coping with changes in technology: the introduction of blade servers is an example where a number of buildings had to have their power supply capacities increased.

■ Coping with changes in legislation and regulations that necessitated upgrading of building services engineering systems due to new performance requirements, either for people or for equipment and processes.

By contrast, design margins cover design contingency, which should be reduced as the design progresses.

System selection and design

The selection of systems starts with considering all feasible options, appraising them on the basis of the most significant factors and then making a final selection. An important consideration in system design is zoning. This is about dividing building areas into different zones so that you can control them at different temperatures or different conditions – whatever might be necessary.

Requirements for utility services

The requirements for capacity and equipment for the public utility services starts with estimates, often based on load densities; for example, per m^2 of floor area, per bedroom (for hotels), per cover (for restaurants) and per bed (for hospitals). During the design phase more data on the load profile is generated: for knowing what energy consuming systems are being provided and understanding, from the end-user how the building will be used.

Typically, in the early stages of design building services engineers make use of published reference figures (benchmarks or rules of thumb) to generate an estimate for the maximum demand size of a utility service.

It is essential to check, at the earliest point, that the anticipated required utility services can be sourced, or the whole project may not be viable or the design would need to be adapted to suit. For example, if gas is not available for cooking or heating, electricity may need to be considered, or if the quantity and rate of flow of water supply is inadequate then storage tanks will need to be sized and provided, and if there is insufficient electricity capacity, standby generators may have to be used to support the base load, rather than just used for standby power.

Rules of thumb figures are usually based on recent practice and experience, for specific conditions, but they may not represent the impact of future trends such as climate changes or changes in legislation. Benchmark figures may be aspirational targets, achieved under a particular set of conditions, and some published figures will contain design margins, spare capacity and contingency factors, while others will not. It is important to understand these details.

If possible, building services engineers should use reference data for similar completed installations. It may be possible to extract information from recent utility bills or to take actual utility consumption figures by means either from the BMS records or by installing temporary measuring equipment.

The provision for backup utility supplies needs consideration. Some requirements are driven by regulatory matters that impose requirements for backup to life safety equipment. Others are the prerogative of clients based on end-user requirements.

Cost estimates

Although design fees are often less than one per cent of the life-cycle cost of a project (or less than 10 per cent of the total construction costs), the design phase is the single most important influence on those costs. This covers capital cost of the installation as well as added value by delivering functionality, quality, enhanced services, reducing whole-life costs, construction time and defects, and delivering wider social and environmental benefits.

The cost estimates at the outset of design will be refined as the design progresses, and allowances for design contingency will reduce. At a point towards the end of the design process, which may or may not be when the design work is absolutely complete, a package of design information will be deemed to be frozen for the purposes of defining a baseline cost. Thereafter, a change control procedure may be introduced to ensure that if the baseline information is changed the associated changes in cost are monitored and tracked.

Compliance with requirements of enforcing authorities

At all stages of the design it is important to ascertain which are the enforcing authorities and what their requirements are and to understand the information that has to be submitted to them for approval.

Design for maintenance

Maintenance is done to ensure that the availability of important assets (the building and its functions), to prevent loss of core business (due to a failure or breakdown in any of the support activities such as the building services engineering systems) and to prevent the excessive cost of catastrophic plant failures and a number of undesirable outcomes, such as customer dissatisfaction, non-compliance with legal requirements, health and safety problems, increase of energy consumption and environmental loss. In these respects building services engineers scope of work requires consideration in the design stage of many different factors, involving system design, performance characteristics, reliability, human factors, safety, logistics, quality, reconfigurability, flexibility, testability, producibility, disposability, environmental considerations and economic factors.

The maintenance strategy and the availability of spares must also be considered at an early stage. Some buildings are associated with industrial sites which may have engineering resources available to maintain and repair equipment, but others are remote from such facilities. Some areas have local stockists and agents that can be relied upon to provide replacements, while others do not. Building services engineers must consider these points and decide whether or not the design can rely on the ability of local maintenance staff using readily replaceable spare parts carried by the building owner, local stockists or replacements that have to come from overseas. In any case, an effective system of maintenance procedures must be established. Comprehensive maintenance manuals must be left with responsible personnel and be written in the language used locally. Effective testing procedures must be established, involving regular test running of plant and equipment, and be properly recorded.

At completion of each stage, building services engineers should provide clients with technical appraisals and recommendations in order to inform the way in which the project is to proceed. This is usually formalised in a report, possibly supplemented by a presentation.

The specification of building services engineering involves complex trade-offs between performance, floor space take-up, capital and whole-life costs, maintenance and renewal strategies and environmental concerns. Different clients have different priorities, which need to be reflected in the design.

The reports at each stage should cover:

■ deliverables in accordance with the deliverables schedule for that stage
■ reporting of any deviations from the original brief
■ schedule of information required from other parties.

At the end of the design phase the information should describe all the building services engineering systems. This includes verifying sizing of plant and equipment, evidenced by supporting calculations and coordinated drawings.

Pre-construction

Once the design has been finalised, the information is used to prepare for the particular tender process, go through the tender process and evaluation, and this culminates in the award of contracts to the construction team.

Tender preparation

Using the design information developed by the end of the design phase, building services engineers should be able to prepare tender documents. There additional non-technical issues which enables those tendering to assess the resources they will need to allocate to fulfil the project requirements. Depending on the form of contract and project, these may need to be considered:

■ Bills of quantities. These itemise, to varying degrees of detail, the materials, equipment and resources which make up the total costs. They will include contingencies to cover unforeseeable costs, such as due to site conditions found after award of contract and changes in legislation. These are usually prepared by the quantity surveyors for review by building services engineers.
■ Scope of works. This is a narrative which describes the total works including building services engineering systems.
■ Preliminaries. These are site overheads that are needed, and includes the provision of temporary utility services for the site works and site accommodation.
■ Identification of any phasing requirements, and define phases, if the project is to be phased. This can be complicated by items that appear in buildings intended for later phases but that are required

for the operation of earlier phases, such as electrical switchgear and runs of pipework.

■ Requirements for mock-ups, testing, samples or models. These may be necessary to satisfy performance or public relations requirements, including computer generated images.

■ Identification of long lead time equipment and systems. It may be necessary to make an early choice of manufacturer, especially if they require specialist input. These could cause sequencing conflicts or could delay the completion and beneficial use of the project.

■ Restricted suppliers list. It may be necessary to restrict the list of potential equipment and system suppliers. This may vary from nominating a specific supplier to specifying a short list to being completely open. The short list may be based on the client's preferred suppliers or as recommended by building services engineers.

■ Deleterious materials. This is a schedule of materials which are either dangerous to health or which are the causes of failures in buildings and, increasingly, materials which are environmentally damaging are included, such as chlorofluorocarbons (CFCs), hydrochlorofluorocarbons (HCFCs) and hydrofluorocarbons (HFCs) are considered to be 'greenhouse' gasses, and lead, found in materials that might be ingested or absorbed such as pipes and paint.

■ Subcontractor pre-qualification. Analyse the scope of work in each trade category and identify appropriate subcontractors who are qualified to perform this work. Subcontractors to be selected from existing pre-qualified database and evaluated on the basis of experience, staffing capability and financial stability.

■ Factory acceptance tests. The aim of factory acceptance tests is to confirm that the requirements for the specification are met, particularly for custom fabricated units such as air-handling units, electrical switchgear and bespoke luminaires. Factory tests allow the stakeholders with the highest level of interest in the outcome to witness the tests. By witnessing the verification testing of critical performance parameters, the designer, client and contractor can feel confident that the special performance features that were purchased have been provided. The manufacturer will also have a vested interest in this test approach because it allows them to document that the unit as shipped met the project specifications and limits their liability for defects that may show up in the field due to improper shipping, handling or installation. The factory often provides a better test setting than the construction site since special instrumentation required is readily available.

■ Requirements of enforcing authorities. Any sign-offs, approvals and agreements for derogations required at this stage should be obtained and documented.

Tender documents

The tender documentation needs to provide sufficient definition of the project to allow for the requirements of the tendering process. The level of detail produced will depend on whether the construction on site will be built in accordance with the information produced by the design team or based on information developed further by the construction team.

In addition, if the design work is not complete, an information release schedule (IRS) may form part of the tender documentation; this provides dates and details for the release of information from the building services engineers.

Tender process

The tender process will vary according to the specific requirements of the project. They involve inviting prospective tenderers to bid for the work. The selection may be from a pre-approved or ad hoc list, chosen because they meet certain minimum standards in general criteria such as financial standing, experience, capability and competence. It may be that a single 'tenderer' is chosen and a contract negotiated with them.

Commonly used methods of tendering are single-stage selective tendering or two-stage selective tendering. The main difference between the two is that in the two-stage process, the contractor becomes involved in the planning of the project at an earlier stage, so the tenders are submitted on the basis of minimal information, and in the second stage the client's team will develop the precise specification in conjunction with the preferred tenderer. This method is favoured in more complex projects, where the contractor may have significant design input.

Tender evaluation

A tender return can be evaluated for different things. It should adopt and observe the key values of fairness, clarity, simplicity and accountability, as well as reinforcing the idea that the apportionment of risk to the party best placed to assess and manage it is fundamental to the success of a project. It is important that building services engineers clarify what aspects they are supposed to review. This may include:

- compliance with tender documents
- personnel and resources proposed
- methodology – how work will be done
- proposed suppliers, subconsultants and subcontractors
- cost – these criteria may be expressed in terms of financial matters, comprising a simple assessment relating to tender sums, or more complex financial evaluation, including consideration of projected costs over the life cycle of the completed project.

It could also address other non-financial factors such as time and proposed methods or levels of capability, or sometimes a mixture of both – collectively referred to as a 'quality/price balance' or 'matrix'.

During the tender evaluation, supplementary questions may be asked of the tenderers. It is usual practice to circulate responses to all tenderers. Once the evaluations are complete, either a single tenderer or a number will further negotiate their position leading to the tender award.

Tender award

The tender award is based on the information in the tender documents as the basis of design, plus any additions, omissions and changes agreed during the tender period to form the contract information.

Construction stage

During the construction stage, the building is built on site – and off-site for prefabricated equipment – in accordance with the programme. The degree of involvement of building services engineers during construction will vary considerably from one project to another and will depend upon the extent of the duties defined in their commissions. Possible roles and responsibilities include:

- discharging ongoing design responsibilities
- addressing new design requirements
- reviewing installation documentation
- responding to queries generated on site, but relating to the design
- inspection of works, on-site or off-site
- witnessing of commissioning of building services engineering systems.

With respect to the provision of information, the construction team may manage an Information Required Schedule. This lists items of information required against dates for issue.

Ongoing design responsibilities

In the case of design-bid-build projects with a single lead design organisation and clear delegations of design responsibilities, building services engineers design responsibility ceases at the tender stage. However, in the case of design and build and PFI projects some design work will carry on during the construction phase. This involves developing the design intent information previously undertaken, until its completion. This may either be done by the original design building services engineering organisation who have been novated to the construction team, or a new building services engineering entity is appointed. The programme will be driven to suit the construction sequence.

Who's in the driving seat now?

It is not ideal for the ad hoc release of design information to drive the construction sequence. This could lead to scenarios such as completed buildings with utility services supplies not being available for the foreseeable future or utility services provisions to buildings being woefully undersized, such that water tankers, diesel generators and gas bottles have to be located outside, to support the buildings. Or ceilings may be completed, but the building services engineering systems within them are not installed, meaning that at some point they will have be demounted to provide access for installation – then put the ceiling back in place. Similarly, plant equipment cannot be put into place as an opening in the building fabric has been closed up. At the construction stage the timing of release of any final design information is driven by the construction sequence.

If not all the information required for construction has been prepared or issued when the tender documentation is issued or the contract executed then an information release schedule (IRS) may be prepared which gives dates for the release of information from the building services engineers.

New design responsibilities

During the construction phase new design requirements may emerge due to:

- changes in legislation, including any needing to be addressed retrospectively
- emergence of new technologies which are deemed necessary to catch up with
- new client's requirements due to changing business requirements, or simply a change of heart
- recognition of errors in the original design.

Reviewing installation documentation

Building services engineers' documents only represent information sufficient for the contractor to begin 'the contractor's required work', which includes the preparation of detailed construction information. Building services engineers' scope may comprise a watching brief, reviewing and commenting, or a full approval role for any or all of:

- contractor's quality plan which advises how the construction phase will deliver building services systems in accordance with the design requirements
- shop drawings, often produced by the manufacturers of equipment or systems (e.g. generator systems), specialist subcontractors or

fabricators (e.g. ductwork and pipework), which are unique to a particular project; including installation details, they also indicate the coordination of that component with other building components to ensure constructability on-site

- installation drawings, which are produced to provide sufficient information to allow the installer to install; these are based on the actual equipment to be used, particularly in terms of size and fixing requirements
- calculations for equipment and distribution system sizing, based on final equipment selections and site-measured distribution routes
- materials submittals based on manufacturer's product information for the materials being provided. These range from the simple material, such as cable trays and pipes, to manufactured assemblies, such as luminaires and radiators, to bespoke assemblies, such as switchgear and AHUs. The information allows for checking the acceptability of the supplier, to compare the performance against the design requirements, giving any intentions for testing. If there are deviations to the design requirements, they enable a view on the acceptability to be taken
- risk assessments which, along with specific risks identified for the construction phase, must acknowledge and deal with the residual risks identified at the design and pre-construction phases
- method statements, which take into account requirements for installation methods and techniques, sequences of installation, procedures for approvals and site safety precautions, all of which contribute to the specific and final details required for procuring and placing the finished work. They detail the way that a work task or process is to be completed include risk considerations to address the particular hazards involved and include a step-by-step guide on how to do the job safely. They should also consider control measures introduced to ensure the safety of anyone who is affected by the task or process
- inspection and test plans (ITP) which record the inspection and testing requirements relevant to a specific building services engineering system or part of a system: to ensure and verify whether the installation has been undertaken to the required standard and requirements, and that records are kept. ITPs may identify by whom and at what stage or frequency particular inspections occur, as well as any specific hold and witness points. They may include references to relevant standards, acceptance criteria and the records to be maintained
- commissioning plans should be in development from the earliest design stages. At the construction stage they should include details of systems tests and procedures, assembly specific checklists, and testing and documentation responsibilities – all related to the design intent

- samples and mock-ups comprise representative portions of the final specified items. The degree of functionality varies, from none to fully functioning. The purpose ranges from reviewing materials, colours and finishes to making final selections from a range of possible items. Mock-ups allow for verifying installation issues. They may be full size or to scale. Samples and mock-ups may be retained as benchmarks for use in project quality control

Caveats

Usually the building services engineers will caveat their comments with something like 'Comments made by the reviewer do not relieve the contractor from compliance with requirements of the contract documents', perhaps adding that the contractor is responsible for all dimensions, site conditions, coordination with other trades and information pertaining to the fabrication process.

Examples of information that would be expected to be found on installation information, but not within tender information are:

- selection of equipment, components, systems and software to meet the specified performance parameters
- preparation of coordinated installation details, as specified, and spatial coordination of the works, based on the design issue layouts and schematics
- coordination of technical requirements and details at interfaces with trade contractors
- design of support systems and any proprietary products supplied
- confirmation of pump duties based on final installation drawings and equipment details
- confirmation of fan duties based on final selection of fan units
- sizing and positioning of refrigerant pipework based on actual site conditions
- sizing of cable containment, including cable trays, ladders, trunking and conduit
- sizing of distribution boards and size of all final circuits from distribution boards in relation to grouping and containment routes
- details of coordination between building services engineering systems, building fabric and structure
- builders' work information for plinths for equipment, and openings for distribution of building services engineering systems
- requirements for secondary steelwork structures and bracketry, fixing systems, lifting beams and access panels for maintenance

- design of syphonic rainwater system, based on measured site conditions
- checking the cable sizing for all sub mains, using selected equipment, actual loads and site measured cable lengths
- checking of acoustic performance and attenuation of plant and equipment based on final installation drawings and equipment details
- developing the design intent details using selected system/ equipment supplier details for lighting control system and lightning protection system
- details of expansion joints in building services engineering systems to take into account building movement
- carrying out electrical grading study of electrical distribution system to prove that discrimination will work
- developing a switchgear interlocking scheme with switchgear vendor
- sprinkler system design and certification

Fitting it all in

So where do all these services actually go? A huge range of things that need to be fitted within a building, and all these have to be coordinated, not only at the design stage, but of equal importance is the actual installation of that equipment. We need to ensure that the design can actually be installed; that it can deliver its design intent; that it can be commissioned – that is, that pipes and ducts have their correct flow rates, that radiators reach their correct temperatures, that grilles deliver the correct volumes of air, and so on. And, finally, that the building can be maintained in the future, for – building services equipment doesn't last forever. At some time in the future, things will need to be replaced, maintained or repaired, and we also require access for that – and this again is important!

Building services engineers will check the installation information for compliance with the contract design intent information. Shop drawings are an integral part of the construction process and present unique legal challenges. Although construction details are produced by contractors and their subcontractors, they may be reviewed by the building services (design) engineers before being installed. Thus, constructors and sub-contractors may venture into the design process and subject themselves to its attendant risks and liabilities.

Building services engineers may review the FF&E equipment information to ensure that the effect on the building services system design are considered, especially as FF&E is often only finally selected after the design phase is completed and assumptions will have been made during the design stage, including, for example:

- requirements for utility services: electricity, gas and water (supply and drainage), in terms of capacity and physical point of connections
- requirements for monitoring and control by BMS systems
- effects on the local environment, such as heat gains which affect the cooling system design, modifications to airflows which affect the HVAC system design.

Responding to site queries

Building services engineers may need to visit or reside on site, full time or part time to resolve design problems. This is particularly relevant when ground conditions or the details of existing installation could not have been evaluated until the contractor had access to the site and had started to excavate or open it up. It may also be practical to be on hand to address issues quickly and reduce paperwork.

Building services engineers may, despite robust QA/QC procedures, make mistakes in their designs; also, not all problems can be foreseen when pen meets paper. When a project runs smoothly, problems discovered during construction are quickly corrected. An on-site building services engineer has the best mental picture of how the project's components relate to each other and how to make those corrections. They should have the experience to spot problems and deviations early on, before they become too expensive or difficult to correct.

Also, from a client's perspective it may be comforting to know that building services engineers are still available to them during the construction phase to provide informed reports of the project's progress, a trained eye toward quality control and a protection against work that is not according to plan. With any building project, the familiar caution holds true: Expect the unexpected. Unanticipated problems – and opportunities – which will arise during the course of construction. With an intimate knowledge of their project's history, on-site building services engineers are valuable assets in seizing new opportunities that are consistent with clients design objectives.

The construction team may raise queries or requests for clarifications which can be categorised as:

- administrative
- technical
- request for deviation from the contract documentation.

Building services engineers need to first reflect as to whether they are contractually in a position to respond. Building services engineers will have to deal with requests from the construction team to deviate from the documentation included in the contract award. This may be for a piece of equipment, system, material type, method of installation, testing criteria or use of a particular supplier. Possible reasons that may be cited include:

- to meet programme requirements for delivery or installation times
- to save money, either capital cost or running costs
- because the originally specified item is no longer available
- because the construction team believes that they can provide a 'better' alternative.

The onus is on the construction team to justify the reasons for the change.

Building services engineers needs to evaluate and approve design changes but must ensure that there is no technical compromise.

Inspection of on-site works

For on-site works this involves checking the installation for compliance with the contract design information. In particular, the inspections should verify that the correct materials have been used. Also they should recognise and review if the installation is different from the design intent, in such a way that the design calculations are voided.

Routing of pipes

Pushing water through a straight pipe is one thing but any additional resistance to that water flow means the water pump has to work harder. And having the water bend round here and there is going to create turbulence and create resistance, which means that the pump has to work harder, and maybe the originally specified pump will no longer be adequate. And what if this is not picked up until the commissioning stage, or the system's up and running, and you're just not getting enough water somewhere? What cost then to put things right? ... and so on and so forth. The whole thing – a small change to the planned routing of a pipe somewhere – has knock-on consequences, often hidden, which is why the more things you can sort out, coordinate and communicate properly, early in the process, the better.

For off-site works factory acceptance tests will ensure that design, operation and maintenance features can be verified prior to the units' shipment, and corrections can be made at the factory. Even with a detailed specification and submittal review process, misinterpretations can occur. Occasionally a special feature does not work out as anticipated. Detecting these problems in the factory allows them to be corrected in a controlled environment by people intimately familiar with the required fabrication processes. Most equipment manufacturers will have a vested interest in this approach, because they avoid the costs associated with sending skilled technicians and engineers out to the site to correct a deficiency not caught by their own quality control process.

Attendance at a factory acceptance test (FAT) will vary according to how the responsibilities of the contract are allocated, but it might include a client's representative, a building services engineer (or their representative), the commissioning services provider and the subcontractor.

Alternatively the responsibility for a FAT may be passed to a third party inspectorate (TPI), but the nature of the responsibilities still needs to be properly understood and to be seamless. Even passing the work to a TPI can cause problems with respect to demarcation of responsibility, proper briefing and follow-up.

During transportation, equipment may be left outside, roughly handled and generally treated in a casual way. During storage, materials are again likely to be exposed to the elements, so any precautions which need to be taken should be arranged. Many items of equipment may have to be inspected or turned over by hand at regular intervals. Control devices, electric motors and starters, control panels and instrumentation are likely to require special consideration. It may be necessary to provide air-conditioned site storage for certain items of equipment.

Typically site acceptance tests (SATs) will be required to re-verify some of the factory test items. This is especially true for larger equipment that must be disassembled for shipment. The responsibility for correcting problems that show up in these re-tests usually falls to the contractor, not the equipment supplier. If the unit was not factory tested, the lines of responsibility for correcting problems uncovered in the field can become controversial, especially if the problems are not immediately discovered.

Other equipment and components may be manufactured off-site, but not inspected; however, there are a plethora of different agencies involved in assuring quality and consistency in components and equipment: type testing organisations, national (and other) accreditation bodies, calibration laboratories, standards testing authorities, underwriting laboratories, certification bodies and so on, which have different reasons for existing and hence different levels of significance and relevance in the testing process.

Accreditation is a formal, third party recognition of competence to perform specific tasks. It provides a means of identifying a proven, competent evaluator so that the selection of a laboratory, inspection or certification body is an informed choice.

> **What is UKAS?**
>
> The United Kingdom Accreditation Service (UKAS) is the only national accreditation body recognised by the government to assess against internationally agreed standards, organisations that provide certification, testing and inspection and calibration services in the UK.
>
> Accreditation by UKAS means that evaluators – testing and calibration laboratories, inspection and certification bodies – have been assessed against internationally recognised standards to demonstrate their competence, impartiality and performance capability.
>
> There are equivalents worldwide and there are multilateral agreements for the purposes of mutual recognition through the European Co-operation for Accreditation (EA), the International Accreditation Forum (IAF) and the International Laboratory Accreditation Cooperation (ILAC). Those bodies that are signatory to these agreements are deemed to be equivalent, having undergone stringent peer evaluations.

Witnessing of commissioning

Commissioning focuses on verifying and documenting that building services engineering systems can be used, operated and maintained to meet the design performance requirements. To ensure that newly installed building services engineering systems will perform properly when put into operation, contractors are required to verify through in situ measurements that the performance of the systems they supplied and installed is in compliance with the specified requirements before the systems can be accepted. Building services engineers' roles may include inspection and witnessing the commissioning.

A building services engineering installation may appear to be well constructed until it moves from its static to its dynamic state (e.g. when water is put in the pipes or the electricity is switched on) when it may fail to perform as expected. At this stage, building services engineering systems may also have to be tested, proven and witnessed to the satisfaction of designers, enforcing bodies, utility providers and insurers, and any delays may have a significant effect on the programme completion. This contrasts with structural engineering, where, although beams deflect and the whole building may settle, in general any major issues are discovered much earlier.

Test methods and validation methodology should be specified by the building services engineers during the design phase. These should be incorporated into a commissioning programme which building services engineers should review. This should incorporate milestones, sequential paths, interrelationships and individual systems, and should detail how the different hand-over phases will be commissioned then recommissioned as more and more gets handed over and brought on line. Where available and appropriate, test methods and validation methodologies should be taken from a standard, code of practice or similar reliable source. Also, during the commissioning, design information is required to be revisited, so it should be readily available. This includes – for example for a ventilation system – design airflow rates, design air velocities, pressure differentials and noise levels.

Handover and close-out

It is only after handover and with the building in operation that the building services engineering systems are truly tested for client expectations, design intent and performance outcomes and functionality. Often operational issues arise when people use and occupy the building – people who are not experts in managing it but, nonetheless, have knowledge and opinions that they wish to vocalise about its performance in relation to their own objectives. A separation between construction and operation can mean that buildings are handed over in a poor state of operational readiness, particularly when programme delays have led to condensed testing and commissioning and pre-handover periods.

Contract administrators certify practical completion when all the works described in the contract have been carried out. Practical completion is referred to as 'substantial completion' on some forms of contract. This has financial significance, but also provides the starting point for the defects liability period.

Handover should involve building services engineers, other members of the design team, the construction team, operators and commissioning and controls specialists, in order to strengthen the operational readiness of the building. This includes verifying that all handover documentation is completed to a satisfactory standard and is adequate. This may include:

■ operations and maintenance (O&M) manuals and as-built drawings provided, checked and approved in accordance with the contract requirements
■ building owner's manual
■ building user's guide
■ health and safety file
■ building log book, with guidance on energy targets and monitoring
■ construction stage report
■ basic design information
■ commissioning results, including detailed testing and inspection (T&I), testing and balancing (TAB) and site testing records, which have been appropriately approved by the commissioning manager in all respects
■ validation report
■ detailed snagging inspection list, which has been undertaken and recorded in sufficient detail such that it is concise and factual, and

has photographs and a short description, and also agreement as to how and when outstanding items will be addressed
- statutory signage in place
- agreed training completed for building operators
- keys and other security codes and passwords.

Is it finished yet?

Determining the point at which a project is complete is not as simple as it seems. In the last stages of construction, building services engineers and everyone else are tired and eager to move on. Contractors may consider a project is complete sooner than others. Building services engineers can estimate the extent of completion against the contract requirements and fairly note any items that remain to be completed. Naturally, there can be tension between a contractor's desire to get paid as much as possible as soon as possible, and the need to ensure that payments are in proper proportion to the work that is completed and that they do not get overpaid or paid until all their obligations to you are fulfilled.

The defects liability period begins upon certification of practical completion and typically lasts between six months and two years, or for a period stated in the original contract. This may be issued with an attached list of minor omissions and defects to be rectified in the defects period. An example of this would be if the certificate of practical completion might trigger tenants fit-out and subsequent payment of rent when it is in nobody's interest to delay the programme just for the delivery of a piece of door furniture or a replacement luminaire.

Once the certificate of practical completion has been issued, the client takes possession of the works for their occupation.

In use

Clients sometimes accept buildings at handover whose building services engineering systems work as per the specifications, but do not work optimally or as expected or have problems not apparent until the building is operational. There may be issues with energy use, equipment reliability, system durability, occupant comfort, worker productivity or environmental performance. Thus, it is necessary for building services engineers to remain involved with buildings beyond practical completion, assisting clients during the first months of operation and beyond, to help fine tune and de-snag the systems, and ensure that the end-users understand how to control and best use their buildings.

What not to expect

So people just don't walk into a building and say: 'Wow, this building is really comfortable; what a lovely working environment!' or even less likely, 'Gosh, those building services engineers did well and somebody should win an award for this!'

It just does not happen. Architects win awards for beautifully designed buildings of glass and steel with attractive staircases and things. But building services engineers? No! The best that building services engineers can hope for, a bit of luck, is an absence of complaints.

Of course, if nobody moans, does that mean that everything's OK? How do we know that we're putting in a really good design, or the best possible design? Or that we are achieving the best possible comfort conditions for that building? Or that this is the best choice for a system? How do we know? What feedback do we get?

I'll tell you now that most consultancies are very reluctant to go back and say to people 'Did you like this; is it working all right; are you happy with it?' Because it's a litigious industry; everybody's frightened that someone might say 'No, and now you come to mention it, we want to have words about this.'

For if there are problems then everybody's looking for somebody to blame. Whose fault was it? What happened? What went wrong? And it can cost millions of pounds to put right.

Of course, there have been studies, but these are few and far between, which is a pity because on the facilities management side, you're in an ideal place to get reliable data. It's the facility management team that have to cope with the day-to-day problems and complaints.

But that still leaves the basic question: how do we know?

And this is still one of the problems with the building services industry. It's a huge part of the profession, but it's mostly hidden, it's under recognised, and people hardly ever say thank you – nobody really ever says, 'Yes this is great.' All you ever hear is either nothing, or lots of complaints.

On the other hand, the plus side is that we don't normally tend to kill people with our designs. Occasionally the odd boiler or pump may go wrong somewhere, but on the whole, we don't make buildings collapse or create major disasters; we just make people feel a little uncomfortable and fidgety.

In essence, the question we are really asking is 'Are you sitting comfortably?' and if we look around a typical work space, what we really notice is that people vary. Some people have shirts or blouses with short sleeves, some long sleeves. Some wear jumpers, others jackets … in other words people tend to adjust their individual dress code to try to find their own personal comfort. But this also raises the question of what can the system deliver; how quickly do they respond to changes and how accurately do we need to measure them? For example, if we have large windows into which we suddenly get a burst of sunshine causing the occupants to get too warm on one side of a building, how quickly should the system respond to that? And that's really a control issue.

So we're trying to cope with things like wind, sunlight and so on, and the fact that people vary. There is a wonderful term which is used when you're looking at systems designed to produce comfort: PPD – percentage of people dissatisfied.

And the problem we face when trying to decide what is an acceptable comfortable temperature for a space – 18, 19, 20, 21, 22, 23 – is that whatever factor you pick, somebody will not be comfortable. So what we aim for is, quite literally, the old adage:

'keep most of the people happy most of the time'.

In the summertime, for example, we might pick a value of 21–23 °C and that should, ideally, give us an 80% satisfaction rating. However, what that means is that 10% of people will be a little too warm and 10% will be too cool. Perhaps they can take a jacket off, and the others put their jacket on, but with any luck most of the people will be mostly all right … and you'll notice the very precise engineering terms we're using here!

Right, so we've selected a range 21–23 °C and a range because, as we've seen, firstly we get a vertical temperature range within the space and secondly because if our controls decide you want a bit more cooling, it will take time for the sensor to tell the system to send more cool air, for this air to get to where it was required and for it to start cooling, during which time the temperature will have risen a bit more. So whatever we do, the temperature is always going to vary by a few degrees at any point in the space and it's also going to vary by several degrees from floor to ceiling anyway.

So we're mostly hoping for something that's going to be acceptable, and to achieve this, we need to look at the range that will actually occur within the space and ensure that we keep reasonably comfortable conditions all the time. That means we therefore need to look at what the various systems can and should deliver and what controls we have that achieve this.

Also, the operational pattern of a building is influenced by the activities within. These may be different from those considered for the design basis and may change over the lifetime of a building. The ability to recommission needs to be integral to the design.

Discourse with clients provides a conduit for feedback methods, particularly on matters such as energy and occupant surveys. This may help with managing any outstanding snagging. Deficiencies that were not identified before occupancy may come to the attention of facility management personnel through user complaints or routine operation. Initial underperformance can easily remain undetected, leading to long-term chronic problems that never get fixed.

During this period, clients may report defects that arise to the contract administrator who decides whether they are defects in the works – things that are not in accordance with the contract – or whether they are maintenance issues. If they are seen to be defects, then instructions are issued to the construction team to make good the defects within a reasonable time.

At the end of the defects liability period the contract administrator prepares a schedule of defects, listing those defects that have not yet been rectified, and agrees with the contractor the date by which they will be rectified. The construction team must in any event rectify defects within a reasonable time.

When the contract administrator considers that all items on the schedule of defects have been rectified, they issue a certificate of making good defects. This has the effect of releasing the remainder of any retention and results in the final certificate being issued.

Notwithstanding this process, defects caused by failures in design, workmanship or materials, may not become apparent or be readily detectable, even with the exercise of reasonable care, until after the end of the defects liability period: these are called latent defects. Examples of common latent defects include:

- a supposedly waterproofed luminaire starting to allow water penetration
- equipment located outdoors, not coping with the local microclimatic conditions, which may have changed since the design stage and weathering to a point that its functionality is affected
- under-strength concrete or misplaced reinforcement allowing movement damage to the structure which shifts building services engineering distribution equipment out of alignment.

Complex arguments arise in respect of economic loss due to latent defects. Commercial fixes for latent defects and successive owners include collateral warranties, guarantees, building warranty schemes and latent defects insurance. Latent defects insurance, also known as decennial insurance, may need to be considered.

9.2   Design management issues

There are a number of issues that embrace the whole design management process, that is, they are not specific to a particular stage of a project. These need to be continually addressed throughout a project.

Design responsibility matrix

A design responsibility matrix (DRM) defines the roles of the various project team members, and shows the connections between work that needs to be done and design team members. This can be used during all project phases and can extend to the likes of suppliers and other parties not necessarily included in the contract.

Setting up a DRM involves determining the activities, the participants (which may be individuals or organisations) and their level of participation, to which all the named parties agree. The level of participation needs to be defined; for example the following could be considered:

■ lead responsibility – who will lead and take final responsibility for the responsibility; there must only one lead for any particular task
■ inputter, whose input is required for a task
■ informed – those who are kept up-to-date on progress, often only on completion of the task or deliverable, and with whom there is just one-way communication (informational only)
■ verifier – those who check whether the responsibility is being correctly carried out.

Despite the straightforward nature of the information included in a DRM, getting all the participants to agree can be challenging.

Hierarchy of legislation and standards

It is important to understand, agree and record the legislative framework for the building, and specific guides and codes. This is particularly important if the project is being undertaken outside the building services engineers' home country, where the same legislation and standards do not necessarily apply and nothing should be assumed.

Building services engineers should be continuously reviewing information provided by professional institutions, local and central government information points and even the trade press for information on any new or changing legislation.

It may be deemed necessary or prudent to derogate or deviate, in part or fully, any of the documents specified. If derogations are required these need to be agreed with the relevant enforcing authority and evidence of this agreement included in the documentation.

Stakeholder analysis

Stakeholder analysis is about identifying all persons, groups, institutions and bodies that may have an interest in a project and taking steps to manage their interests and expectations. They may be affected by the building, both positively or negatively, directly or indirectly. They may or may not have a contractual link to the project.

In general, a stakeholder is anyone who will make use of, develop or have an impact on any aspect of a project. Stakeholders can be either direct or indirect. Direct stakeholders are those people (developers, managers, customers) whose actions can directly impact projects – they are involved in the project life cycle, or are impacted by the project – they use the system or the output that the project puts in place. Indirect stakeholders are those who have some authority to influence the project or those who are interested in its outcomes: stakeholders are those who have a stake in the project.

This analysis needs to be started in the early stages of a project and updated on a rolling basis so that any risks and required communication can be included in the overall project plan. Decisions need to make about how to deal with stakeholders depending on their level of interest and their power to influence project outcomes.

Site visits

It is essential, at the outset, to establish both the existing site conditions and anticipate any potential changes (either on the site or nearby) that might affect the building services engineering design. This process should be ongoing until handover. An appreciation of the physical apparatus is necessary as it may cause hazards and constraints on either an existing or a new site. The emphasis will be different for a new building on a greenfield site to an internal refurbishment of an existing building. However, the principles will be the same and site visits should be undertaken prior to any design work commencing and, depending on the terms of the appointment, during the construction works.

Prior to any design work commencing, all available record information pertaining to the site should be obtained. Sources will include client-held record information, utility services providers and local authorities, for example street lighting. This should be reviewed as part of a desktop study, but it cannot be fully relied upon for any of the following reasons:

- The position of reference points, for example the kerb line, may have changed since the plans were drawn.
- The surface levels may have changed, meaning that the depths shown are now incorrect.
- Services (particularly cables) may have been moved by other parties without the knowledge of their owners/operators.
- Services marked as straight lines may, in practice, snake. In particular, excessively long cables may have been laid in horizontal loops outside substations or switch rooms.
- Plans may show spare ducts which may be either 'used up' or not in a fit state to be used, for example due to ingress of tree roots, water damage, collapse or subsidence.
- The routes of older services in particular may not have been recorded in the first place.

The focus of initial site visits is to verify the record information. This will identify any gaps and ambiguities and will determine the need for arrangements of and definition of the extent of special investigations or tests (either intrusive or non-intrusive).

Parts of building services engineering systems are buried in the ground, or in inaccessible voids or risers, or are embedded in the building fabric making it particularly challenging to verify their nature, condition and exact location by observation.

Health and safety responsibilities

Building services engineers are in a position to influence and reduce the risks that arise during construction, and also during use, maintenance, refurbishment and eventually demolition. Decisions taken by building services engineers fundamentally affect the risks faced by workers and those who use the building.

Health and safety must be an integral part of the design process, not an afterthought. By including considerations of health and safety in the design management of the project, hazards can be identified early on, so they can be eliminated or reduced at the design stage, and the remaining risks can be properly managed. It does require knowledge of construction processes, and it needs to take into account unforeseeable hazards or exercise any health and safety management functions over contractors or other designers. Their primary responsibility is to minimise hazards and risks to people by good design. Initially they should try to eliminate hazards. If hazards cannot be eliminated then risks should be minimised and designers should provide information on any significant residual risk. This should be managed through structured hazard identification throughout the building's life cycle.

Life safety compliant designs

Ensure that the building services engineering designs are in accordance with good practice as well as applicable building, fire and safety regulations. Some aspects of life safety design, in some jurisdictions, will be subject to checking and ultimate sign-off by an authority independent of the building services engineering entity. In others, it is deemed that building services engineers will have done the appropriate due diligence. In these situations it is vital that someone with the highest expert skill, rather than just a competent person, has the final responsibility for checking the design.

Avoiding working at height

Working at height is an inevitable hazard not only during construction but also during maintenance. However, it may be possible to alleviate it. In the first instance, consider whether any equipment can be located at a lower level; for example, instead of locating luminaires at the top of a high space, use wall-mounted uplighters, so that both installation and maintenance are safer. Also consider the viability of fabrication off-site and lifting into position.

Provide adequate space for plant access operation, maintenance and repair

Space and the cost of providing space for plant and building services engineering distribution is often at a premium. Pressure to reduce the spatial requirements for building services engineering systems installations is therefore an understandable element of the overall design process. However, considerable care is required if building services are to be designed and installed to provide adequate space for the safe and efficient maintenance of the systems.

The main division of tasks is between operation and monitoring, inspection, servicing and maintenance. Operation and monitoring usually require no more than observation and operation of the systems controls. Inspection, servicing and maintenance involves in-situ work with or without dismantling, and/or demounting/removal for on- or off-site works. Demounting requires more local space, and local working requires appropriate areas.

Even when safe access is provided, access to component parts can be so restricted that the cost of maintaining these parts imposes a significant additional operating cost on the maintenance of the building, for example if part of the system or plant room enclosure has to be dismantled.

Other health and safety considerations

Allowances in the design stage must be made for other hazards:

- hot or cold surfaces should be insulated if they might be touched
- sharp edges and ends, usually occurring on equipment, should be smoothed out
- noise – acoustic treatment to the whole of a plant room occupies significant space, so in the case of generator rooms, which can be extremely noisy, check the sounds levels against the quality of ear defenders
- hazardous atmospheres containing, for example, explosive dusts or vapours, petroleum vapours or methane will need special electrical accessories and lighting to be specified
- unprotected moving parts must be identified as a no-go area, and delineation may be required in addition to access space and should not encroach on any occupied or circulation areas
- the need to work on live electrical parts should be designed out as far as possible, even if it means disruption to building occupants
- heavy and/or awkwardly shaped objects may require lifting eyes and/or runways suitable for standard lifting techniques positioned above equipment. The full height of the plant space below the eye or runway must remain free of other services
- falling from heights – walkways, protected ladders and so on must connect with access routes and must not obstruct any other access
- tripping – provision of step-overs for floor-mounted pipes and cables should be coordinated with the layout of services overhead.

Provision of warning signs should be made for escape routes, certain electrical rooms and rooms containing dangerous materials or liquids.

Life cycle considerations

Building services engineering systems are typically designed to function in a building for a lifespan of 20–25 years as a maximum, but often the rate of change of use of areas, changing technologies and statutory requirements, means that this time span is very much less – particularly for ICT based systems. This contrasts with the structural engineering solutions which are usually designed for a much longer lifespan. The building services engineering services installations are adapted more often and should be designed accordingly.

Building services engineering systems may be more difficult to access. Pipework, for example, might be buried within the floor so that if it needs replacing, the whole floor has to be dug up. Hardly anything within building services has a lifespan longer than about 25 years, and many items a lot shorter. In equipment such as boilers or cooling plant or cold water tanks, metal can corrode and rust. Pumps and their motors might need replacing every five years or so, which means that many items are going to have to be replaced several times during the life of a building.

Lessons to learn

When you teach architectural students, there are three really awkward questions you can ask them. The first is 'How do you get access for the fire brigade?', the second, 'How do you get rid of the rubbish?' and the third 'How do you clean the windows?'

I've had architects tell me, 'Well, they'll carry the rubbish down in the lift', and again 'The fire brigade can also use the lift', or 'Cleaning windows of this nice curved roof atrium? No problem, we'll use self-cleaning glass.'

But another equally important question today is 'How do we get access to replace the building services?' You can't really dismantle it and take it down in a lift – it is just too big. So in terms of replacing some items of plant, you may have to think about access to the whole building. This could mean, for example, bringing a crane up next to your building to lift out a boiler. Think of a building like The Shard or other high-rise buildings. If you actually look at them, one of the interesting things you'll notice is that every 10–15 floors or so you've got a floor that looks a bit different. Sometimes it's still covered with a skin and so is difficult to spot, but with other buildings you can see louvres and grilles – for the very good reason that if we had just one large plant room on the top of a 50-storey building, it and the pipes and ducts leading to and from it, would be absolutely huge and totally impractical.

So for buildings which have lots and lots of floors, it is quite common to have several intermediate plant rooms, say every 10 to 15 floors, just so that any one item of plant and one system of distribution is not impossibly large.

But either way, whether you've got equipment right at the top, right at the bottom or it's located on intermediate floors, you will still need, at some time, to get in and replace it.

So just how do you replace an item as large as a commercial boiler, a chiller or an air-conditioning plant? With something small like a pump, it might be small enough to put it into a goods lift, but with a big item of plant, you may need to take a side off the building! And what else does that entail? You will probably have to close off a street and ensure access for a crane … all of which will be much easier if they are thought about at the design stage of the project.

It makes more sense, therefore, to have the main access to your plant room in a side street, so if you do have to close off a street, you're not inconveniencing too many people. It may even be one of the planning requirements.

Managing FF&E requirements

The inclusion of a client's FF&E needs to be managed throughout the process. FF&E may be either procured separately or as part of the construction contract, or a mixture of both. However, their requirements may need to be allowed for in the design process, for example:

- built-in or standalone equipment requiring connections to the building services systems: power, air, water, drainage and telecommunications connection or that affects the internal environmental conditions, such as heat gains
- other built-in or standalone equipment that does not require building services systems connections such as cupboards or shelving but whose location may impact on the position of socket outlets, grilles, taps, telephone and so on; these should be described in terms of their location and size.

Areas of potential overlapping responsibilities

There are areas of design where there is potential for building services engineers' responsibilities to overlap with the work of other members of the design team. It may be that one entity makes an assumption that another party is undertaking an area of work. Typically two or more parties need to contribute to a particular area of design and the real issue is who should take the lead coordination role. This provides the potential for either conflict, or for both parties missing out the work (either opportunistically or genuinely) by assuming the other party has covered it. Some potential areas of conflict are interdisciplinary to the building services engineers, for example between mechanical and electrical disciplines. This is particularly important to address when separate organisations are providing the mechanical, electrical or public health disciplines of building services engineering.

Demarcation lines can also provide areas for potentially conflicting responsibilities. The demarcation point determines where the responsibility for the design changes from one entity to another. However, it is not a straightforward case of interconnecting apparatus at that point as the whole distribution network must be designed as a complete system to ensure that end-to-end continuity of design parameters.

There is potential for a situation when all the members of the design team assume that particular responsibilities have been passed to the construction team. This is a slightly different scenario. These occurrences might be acceptable, as long as they concur with the project specific design responsibility matrix.

There may be no definitive 'right' answer as to which party should take the lead responsibility, just issues that need to be addressed on a project-by-project basis as part of the design management process. The following examples describe typical scenarios and explain why these misunderstandings may rise – but it does not attempt to direct the reader to a specific answer.

Building management systems

Electrical building services engineers assume that mechanical building services engineers have the lead responsibility for the design and specification of the building control systems: while mechanical building services engineers assume that the electrical building services engineers are taking the lead.

Building controls systems involve controlling and monitoring both mechanical and electrical equipment. Monitoring and controlling devices are connected by cables and the associated cable containment system either directly, or via a series of intermediary points, connects back to monitoring and control units.

Mechanical control centre panels

Electrical building services engineers assume the mechanical building services engineers have lead responsibility of the design and specification of the mechanical control panels, while mechanical building services engineers assume that the electrical building services engineers are taking the lead.

Mechanical control centre panels are a form of electrical switchgear which may also accommodate motor starters and drives required for the mechanical plant.

Builders' work ducts

Building services engineers assume that structural engineers will provide all the details for builders' work ducts, for example supply air ducts formed by the building structure and fabric, while structural engineers assume that building services engineers will provide the required details.

Builders' work enclosures are not necessarily just about the provision of clear open voids. They may include cast-in fixings for pipes, ductwork and cable containment. The internal finishes need to be considered as they impose a frictional resistance to the air path. One party needs to take the lead in ensuring that these are correctly specified and correctly located.

Secondary steelwork for building services equipment

Building services engineers assume that structural engineers will provide all the details for secondary steelwork required to support building services engineering equipment, while structural engineers assume that building services engineers will provide the required details.

The final design of secondary steelwork is usually undertaken by a specialist subcontractor, who is usually also the installer. However, secondary steelwork needs to be indicated graphically in the design information, but it does not necessarily need to be sized or engineered. This ensures that the construction team is aware that it needs to be included in their costing and planning. The reason for the possible conflict is that the secondary steelwork needs to be fixed to non-structural

walls and ceilings where the stud framing itself is not structurally sufficient to carry the load, so secondary steel in this case is constructed within
non-structural walls and ceilings to supplement its structural capacity.

Design of the building fabric

Building services engineers assume that architects – or sustainability consultants,
building physicists or even code assessors – are taking the lead in specifying the
properties of the building fabric. This includes taking into account its properties
relating to air infiltration, thermal performance, moisture control, contribution
to daylighting and acoustic performance. But the other parties are all assuming
that another party is taking the lead responsibility.

The design of building fabric and selection of materials needs to ensure
their fitness for purpose in terms of flexibility and robustness, and that
they have the required thermal performance to achieve the desired internal design conditions. Failure to optimise the requirements of building
services engineers may increase demand on the building services engineering systems for heating and cooling systems and so decrease the
energy efficiency of buildings. However, the building fabric may be significant to the appearance of the building or may itself be fundamental to
the sustainability concept – for example photovoltaic curtain walling.

Electronic door locks

Building services engineers assume that architects or interior designers have
allowed for the electrical equipment associated with electronic door locks. These
are magnets, solenoids or motors and the associated cableways within the door
frame to open and close doors, and also, detentes, which are associated with door-
hold-open devices. Hence, building services engineers only allow for an electrical
power supply and assume that the rest is designed and specified by others.

The final door and door frame combination will need to contain all
the components to ensure that it functions properly.

Sanitaryware

Building services engineers assume that the design and specification of
sanitaryware, including the sinks, baths, toilets and urinals, and also the taps,
shower attachments, are within the architects or interior designers scope of
work, and this may also include electronic sensing devices for turning on or off
water supplies, or for flushing – but architects that assume the building services
engineers are specifying all the sanitaryware.

Architects may be interested in the aesthetic appearance of sanitaryware in some areas, but not in others. Taps need to be coordinated with
sinks to prevent nuisance splashing. Electrical power supplies and
possibly connections to a BMS may be required for sensing devices.

Lightning protection system

Building services engineers assume that the structural engineers have designed
and specified either the steel columns and/or the reinforcement to act as down
conductors for the lightning protection system and that provision has been

made for connection points for the earthing system, while structural engineers have not made any specific provision.

If steelwork, comprising steel columns and reinforcement, is not specified to be used as down conductors it will not be designed, installed or tested for this use, and it may be difficult or even impossible to retro-fit an installation afterwards – similarly with connection points for earthing. There may also be costs associated with necessary alterations.

Fire stopping details

Building services engineers assume that at the fire compartment boundaries the detailing will be done by architects or fire engineers, while others assume that building services engineers will provide the details for the building services engineering systems.

Fire stopping details provide information on the materials and methodology for ensuring that the boundaries between fire compartments do not compromise the integrity of lines of fire compartmentation.

Lift car lighting

Building services engineers assume that the lighting inside the lift car is designed and specified by architects, interior designs or vertical transportation consultant, while the others assume that building services engineers will provide the information.

Lift cars require a lighting system, including emergency standby lighting, to suit the particular lift. This may be part of the interior design scheme or may be purely functional.

Electrical accessories in furniture

Building services engineers assume that the detailing for small power, data and telephone outlets, including the associated cables and cableways, within fixed furniture, such as office desks, reception desks, benching and seating, is done by the architects, interior designers or FF&E specialists, while the other parties assume that building services engineers will incorporate the details in their deliverables.

The presentation of final outlets is important to consider for the convenience of end-users to plug in and position their portable devices for use. Also, poor design can lead to potentially dangerous trailing cables. The way in which electrical accessories are fitted into furniture can impact the viability and convenience and flexibility of furniture layouts.

Architectural lighting

Building services engineers assume that any specialist bespoke luminaires are fully designed by architects, interior designers or specialist lighting consultants, and this includes the consideration of cables and cableways within the fitting, photometric performance and EMC issues, while the others assume that they will design the visual aspects of luminaires and that the building services engineers will deal with other issues.

Luminaires must be code compliant with respect to electrical safety, including details of cable and cable ways, and EMC issues. The heat gains from the luminaire are needed to size the building's cooling loads. The photometric characteristics of the luminaire are needed for the purpose of undertaking lighting calculations to determine illuminance levels and glare indices.

Reflected ceiling plans

Building services engineers assume that architects or interior designers are taking the lead on producing a coordinated reflected ceiling plans (RCP), while the architects and interior designers assume that the building services engineers are taking the lead.

RCPs show the items located on the ceiling of a room or space. It is referred to as a reflected ceiling plan since it is drawn to display a view of the ceiling as if it were reflected in a mirror on the floor. Different disciplines within building services engineers will have equipment to be located within the ceilings, such as:

- mechanical – grilles, diffusers, louvres
- electrical – luminaires, lighting control sensors, automatic fire detectors, sounders, public address and voice alarm speakers, CCTV cameras
- public health – sprinkler heads
- interior designers – luminaires, connection points for hanging decorations.

Some of the locations for devices will be subject to rules and regulations; for example, smoke detectors cannot be too close to a supply air grille, or any smoke will be blown away before it reaches the detectors. With the others it is case of coordinating and compromising. In some areas, the layout of the ceiling is visually significant, in others it is not.

Internal and external drainage systems

Public heath building services engineers assume that the civil engineers are responsible for the design of the drainage system at the boundary of the building, while civil engineers assume that their responsibility starts at the first manhole.

The internal drainage systems connect with the outside drainage systems at some point to form a continuous system. The demarcation point could, for example, be the boundary of the building, or 1 m, 2 m or 3 m from the building boundary, the first manhole or the site boundary – but the demarcation point needs to be agreed to avoid potential gaps or overlaps.

Multiple building services engineers on the same project

Consider the scenario where there are multiple buildings on the same site with different building services engineering entities responsible for the design of each building, or one building subdivided into smaller projects, such as different floors, tenancies or departments, again with different building services engineering entities responsible for the design of each section. There needs to be a single entity with ownership of the design criteria for those building services systems across the whole project.

Clients should ensure that common standards codes and guidance are referred to, and where there are subjective elements, that common assumptions are used. Where site-wide or building-wide systems are present, the design criteria and parameters used for any section need to recognise the overall site/building design, that is the subsections cannot be designed in insolation. In the case of systems (e.g. sprinkler or fire detection and alarm) where statutory approvals are required it may also be prudent to ensure that common drawing, design software and calculation standards are used.

Electrical distribution system

The whole electrical distribution network across a building or site-wide campus must be designed as a complete system. This includes study and analysis of discrimination, short circuits and fault levels, earth faults, electrical loads, harmonics and voltage drops. Collectively this will demonstrate that:

- the electrical equipment is satisfactorily protected by the system protection equipment (e.g. circuit breakers), particularly in terms of disconnection times and energy let-through. The short-circuit element of the electrical network analysis is used to determine the level of short-circuit current at specific parts of the electrical system
- the protective devices have compatible characteristics to ensure that protective equipment closest to a fault operates first and within code-compliant disconnection times
- electrical equipment that is not involved with the fault remains in service
- automatic disconnection of supply will occur in the event of a fault to earth for the protection against electric shock.

Without an overall responsibility it is impossible to properly design the switchgear and interconnecting cabling. Switchgear may be 'overstressed', that is the potential fault energy of the electrical system – such as from a short circuit – at the switchgear location exceeds the fault energy rating of the switchgear. When it is operated under fault conditions it is unable to cope with the resulting electrical and thermal stress and this can sometimes lead to catastrophic failure: total destruction of the switchgear. Such failures are accompanied by arc discharge products, burning gas clouds and oil mist (in oil switchgear). These can envelop anyone near the switchgear, resulting in serious burn injuries and often death.

Similarly the volt drop allowed for the whole system needs to be apportioned across subsections of a project. Building services engineers should not start their design without understanding and agreeing to the volt drop allowance for 'their' section.

Use of software

Software is used in all phases of projects to model situations and to provide estimates and approximate solutions. The choice of software may be determined by an enforcing authority, as per client's requirements or it may be selected by building services engineers. This may be an open choice or it may be constrained by the needs to interface with software used by other members of the design or construction team.

Building services engineers need to appreciate who the software is provided by and how this may need to managed. The software could be:

- in-house developed
- from manufacturers, which will be biased towards their own equipment
- from professional institutions and other not-for-profit organisations
- from commercial organisations
- 'freebies' of unknown pedigree, meaning, in worst case scenarios that they may be created by a well-meaning enthusiastic but incompetent amateur, or worse still a 'troll'. But they may also be provided by totally proficient people, with an altruistic desire to share their knowledge.

It is important to appreciate whether the organisation has third party accreditation for its software development processes. Some software is endorsed by third parties, possibly with caveats about reliance, but it does not follow that the software is tested and verified by an independent third party with the pedigree to undertake such an exercise. Notwithstanding this, it is prudent to undertake a 'sense' check and to recognise the 'rubbish-in, rubbish-out' scenario.

When setting up a model, building services engineers have to make numerous numerical and physical assumptions. These will depend on a number of factors, such as the scenario modelled, the resources available and the user expertise. Also, all software has its limitations; it is very important to be aware of these when assessing the conclusions drawn from any predictions.

It is good practice to keep records of all the inputs and outputs involved and, in any reports, record the name and release version of the software used.

Summary

Design management is needed wherever there is design going on. This may be by the designers in a design team or those parts of the construction team where design is occurring. The role of the building services design manager is firstly to understand the requirements and structure of the particular contract including any nuances, and then to establish a framework within which the tasks and objectives are kept in focus as the design moves through its stages of development, and finally provide a point of contact on all design management issues. To be successful, building services design managers need to:

- ensure that the appropriate procedures and documentation are in place
- monitor the deliverables and manage information flows
- allow building services engineers sufficient time for coordinating with the stakeholders, including time for reflection
- ensure that building services engineers have the relevant experience and ability for the particular project
- encourage and provide the support to building services engineers to enable them to find solutions to the problems
- provide access to the client for review and the provision of more information
- help building services engineers to understand the full implications of changes and the possible need to re-enter the design cycle.

Building services engineers' detailed design and tender information cannot be used for construction; their documents cannot be passed to contractors for them to start installing from the information contained therein. These documents are not issued for the purpose of construction per se, but instead they are issued to facilitate construction by expressing the design concept. The documents do not contain sufficient information to construct the project, and much more information is required before the work can be done.

10 Risk management

Risks are innate in any building project. The geographical dispersion, significant number of players, technical changeability, technical complexity and large number of inputs are some of the variables that make construction projects challenging. As a result of these interconnections the number and type of risks tend to be significant.

Risk management involves implementing a logical process to identify, evaluate and quantify, share, manage and monitor potential hazards which may substantially affect project activities or the project as a whole. If any of these risks are realised by loss or damage, depending upon the particular circumstances measures may be sought by the injured party from the party responsible. A risk management process aims to reduce risk to an acceptable level.

The ownership of the risk management usually changes as projects progress. Starting with clients it is passed to the design team during the design phases and on to the construction team during construction. Residual risks are passed back to clients after handover when the building is in use.

This is often done using a risk register, which acts as a central repository for all risks identified for a particular project and, for each risk, includes information such as risk probability, impact, countermeasures and risk owner. Typically a risk register contains:

- a description of the risk, based on experience and lessons learnt
- the impact, should this event actually occur, in terms of time, cost and quality and any other KPIs pertinent to a project

Building Services Design Management, First Edition. Jackie Portman.
© 2014 John Wiley & Sons, Ltd. Published 2014 by John Wiley & Sons, Ltd.

○ the impact of a risk can be measured as the likelihood of a specific unwanted event and its unwanted consequences of loss, simply put:

$$impact\ of\ risk = likelihood \times consequence$$

■ a summary of the planned response should the event occur
■ a summary of the mitigation (the actions taken in advance to reduce the probability and/or impact of the event).

The risks are often ranked by risk score so as to highlight the highest priority risks to all involved.

Building services engineers will be concerned with to the risk management process at the project level and also at their own organisational level. While this book concentrates on risk management from the project perspective, there are some overlapping issues. For example, if building services engineering entities start a project without understanding their scope of works and ensuring that they have sufficient fee to cover it they risk financial loss, disputes, claims and compensation – and maybe a bad reputation.

Risk identification

The identification, consisting of recognising and recording hazards, is complicated; each stakeholder will have a different perspective and will present risks to other parties. Risks can be classified in relation to their locus of action. There are various types of risk. This book will consider project risks which are within the project boundary, from the perspective of building services engineers, that could affect the delivery of the business outcome that the project is set up to deliver. Other risks categories are:

■ business risks which affect the operation of the business outcome once it has been delivered by the project
■ environmental risks which are external to the project environment but which nevertheless can affect the project objectives
■ 'force majeure' risks which are exceptional, unforeseen events or circumstances that are beyond the reasonable control of the parties to a contract
■ reputation risks, whereby the status and standing of an entity is put at risk.

Building services engineers need to also consider how their designs may impart risks related to any temporary works and to the operation, maintenance, cleaning, alteration or demolition of the completed building, and also related to inherent risks in the site or its surroundings.

There are certain generic risks that apply to almost all projects and all disciplines: risk of losing electronically stored information, risk of key staff leaving or being ill, dealing with client-initiated variations, economic circumstances and so on. However, there are some risks pertinent to building services engineering design, some of which are included in Tables 10.1–10.5. These comprise risks at a project level, that is those risks affecting the building services engineers' ability to deliver their service for the benefit of a project. They do not cover risks at the individual or organisational level, or the risks directly affecting clients or other members of the design team or construction team.

Table 10.1 Building services engineering design risks – preparation stage.

Risk factor	Effect on building services engineering design
Building services engineer not involved at this stage	Key issues, affecting the fundamental viability of the scheme may not be considered, e.g. extreme external conditions, local pollution and air quality, unavailability or prohibitive costs of providing utility services to a site.
Existing utility services on the site	There may be wayleaves and covenants associated with the incumbent utility services which may impose constraints regarding access and may impact the effect of adding a new building near them. If they need to be moved this may have a significant effect on the project.
Building has a listed building consent	There will be limitations on where and how the building services engineering systems can be installed.
The building is unique	Unique buildings may not fall under the rules and regulations for standard buildings, meaning that statutory approvals are non-standard.

Table 10.2 Building services engineering design risks – design stage.

Risk factor	Effect on building services engineering design
Changes in statutory and legal requirements	These risks are normally out of the control of all the project stakeholders, but building services engineers should always adopt the strategies of monitoring information from professional institutions and trade publications for hints of any changes, especially those that may be implemented retrospectively or have a set changeover date for implementation.
Challenging programme for design works	As the building services engineering design tends to follow on from the architecture and civil and structural engineering, building services engineers will have less slack than others to work within. This may stifle innovation and hinder detailed checking and reviewing.
Reliability of third party information not known	The longer the design goes on without verifying the reliability of such information the longer the risk has to be carried, and the cost contingency for carrying the risk.
Information regarding operation of the building missing	There are certain matters pertaining to the use of the building that should be provided by clients in their capacity as end-user. If building services engineers have to guess, risks are built up, e.g. occupancy of spaces and details of FF&E equipment.

Table 10.3 Building services engineering design risks – construction stage.

Risk factor	Effect on building services engineering design
Installation does not proceed in accordance with the design information	Integrity of design is compromised, maybe invalidating it.
Alternative suppliers used for equipment and components	Integrity of design is compromised, maybe invalidating it as calculations may no longer be valid.
New findings on site	As previously covered areas are uncovered, e.g. buried in the ground externally or hidden in the building fabric indoors, assumptions made the design may not be justifiable.

Table 10.4 Generic risks – handover stage.

Risk factor	Effect on building services engineering design
Insufficient training of end-users on system use	Building services engineering systems performance not optimised as intended in the design.

Table 10.5 Generic risks – In use stage.

Risk factor	Effect on building services engineering design
End-users do not appreciate constraints of design	End-users do not appreciate that on extreme weather days, the internal conditions may not be ideal, and they may need to make adjustments, e.g. warmer or cooler clothing.

Information produced as a consequence of design risk management should be included in tender documents so that contractors (or other designers) can take them into account when pricing their tenders and planning their work.

Risk evaluation and quantification

Once the potential risks have been identified they need to be evaluated and quantified. This involves subjective decisions. Also, the likelihood and impact changes as the project progresses, so the risk register should be updated.

Risk sharing, managing and monitoring

Deciding which courses of action to pursue – risk response strategy – is largely based on the results of risk quantification. Some of the risks identified above can be quantified, controlled or minimised through the risk management process to estimate or assess the likely outcomes or impacts of risks under consideration in case they materialise. Many hazards are potential or latent and their timescales may be unpredictable. A hazardous situation that becomes effective can cause an incident, an accident or a disaster.

Common strategies to manage risks are:

- avoid risks
- reduce risks
- transfer risks
- share the risk.

Sharing the risk

Regarding risk allocation, the concept of 'limitation of liability' dates back more than three hundred years, when the British Parliament declared, as part of maritime law, that a ship's owner should not bear greater liability than the value of the ship's hull (Zoino, 1989). Zaghloul and Hartman (2003) proffered that in this context, every contract allocates risk, but that not all contracts allocate risk equitably or such that the power and authority to manage the risk is allocated along with the risk itself. Hayden and Parsloe (2006) note that ambiguity over design activities can lead to project delays, increased contractual claims and litigation.

Summary

The complex nature of the entire process associated with delivering a finished building demands a robust risk management process to run in parallel with the stages of projects. Some risks are common to all projects, while others are particular to projects.

References

Hayden, G. and Parsloe, C. (1996) *Value Engineering of Building Services*. Bracknell, UK: BSRIA.

Zaghloul, R. and Hartman, F. (2003). Construction contracts: the cost of mistrust. *International Journal of Project Management*, 21(6), pp. 419–424.

Zoino, W. (1989) Cautious risk taking. *Civil Engineering*, 59(10), pp. 65–68.

11 Information management

Information management is concerned with ensuring that the right information is available when required in the right format. This involves implementing the necessary tools and practices for the collection and management of information from one or multiple sources and the distribution of that information to the relevant parties. The emphasis will vary from project inception to handover and beyond.

Building services engineers will be concerned with information management systems within their own entities (such as accounting systems, internal project reporting and correspondence) and the information management systems which interface with external stakeholders at a project level. This book will focus on the external information management issues.

Information management involves a wide and diverse range of types of information: written to verbal, high level master planning level to minute detail, project-specific to background reference, technical to administrative. There are many drivers to integrate the various aspects of information management into common platforms and systems. However, it is still useful, before considering the technology, to consider the components separately. At the highest level these can broadly be categorised as project related information, reference information and knowledge management.

Building Services Design Management, First Edition. Jackie Portman.
© 2014 John Wiley & Sons, Ltd. Published 2014 by John Wiley & Sons, Ltd.

Project related information

The information pertaining to projects that needs to managed consists of documents (drawings, reports, calculations, samples, pictures etc.), correspondence and tracking project processes. This covers information produced by the stakeholders (client, design team, construction team, utility services and enforcing authority) and also those with no contractual links to a project.

An understanding of why and by whom the documents were created, what they are for and how they are stored and retrieved is vital to mitigate misunderstandings and mistakes. This includes primary information (reports, calculations, drawings) and the secondary correspondence relating to them.

Reference information

Reference information comprises material which may be referred to for data and facts either for guidance or as a statutory requirement to be inputted to the design. It may be in hard or soft copy format. Examples include:

- standards for plant rooms, produced by utility services providers
- historical weather data
- maps.

The documentation must always quote the source of reference data to enable readers to consult the original source independently for verification.

Knowledge management

Information management also contributes to the knowledge management processes. This can mitigate waste, loss and inefficient use of knowledge which can lead to inefficiencies and reworking of past problems – including the reinvention of the wheel syndrome. This is straightforward for the easily captured, codified and stored, explicit aspects of knowledge. The danger is that ICT-driven knowledge management strategies may end up objectifying and calcifying knowledge into static, inert information, thus disregarding altogether the role of tacit knowledge that resides within people.

A method is needed which recognises the information that resides in people, not in machines or documents. Social computing tools (such as blogs and wikis) have developed to provide a more unstructured, self-governing approach to the transfer, capture and creation of knowledge through the development of new forms of community, network or matrix. However, such tools for the most part are still based on text and code, and thus represent explicit knowledge transfer. These tools face challenges in distilling meaningful reusable knowledge and intelligible information and ensuring that their content is transmissible through diverse channels, platforms and forums.

Information management systems need to address how the information is generated, stored, retrieved, transferred and read. Also how any security issues are addressed.

With information management it is important to determine and formalise standard methods and protocols to describe how information is structured, how it will be produced and how it will be managed and exchanged. At the most basic level information may be hand written, manually filed, manually retrieved, posted and read from the paper it's printed on. For very small projects this may still be acceptable, but for anything more, ICT systems are used to support the information management process. This will either consist of a document management system or a whole environment.

With a document management system, the information is created outside the system, and the system allows for uploading and distributing information. This may include the provision for manually or automatically naming, creating revisions, distributing tracking, archiving and retrieving and sharing drawings, photographs and documents. It should also include functions for searching and reporting, and ultimately archiving. This would include correspondence between parties, information requests and responses, online applications to mark up and make comments and to manage and report on project-related communications.

A particular case of an information system where information is created within the system is that of building information modelling (BIM). BIM is an ambiguous term that means different things to different professionals. However, most of the proposed definitions align on BIM representing the process of development and use of a computer-generated model to simulate the planning, design, construction and operation of a facility. The resulting model is a data-rich, object-oriented, intelligent and parametric digital representation of the facility, from which views and data appropriate to various users' needs can be extracted and analysed to generate information that can be used to make decisions and to improve the process of delivering the facility.

BIMs are created from a series of three-dimensional objects. Each object is defined only once and then placed in the model in multiple locations as required. The definition of each object may include geometrical information and information describing other properties, such as its materials, construction process, time-related information (such as delivery times) and operational information. If the object is then changed, these changes will appear throughout the model. This automatically makes models consistent and reduces errors. Objects can be defined parametrically, allowing them to be related to other objects, so that, for example, they might have a common colour, which when changed on one will also change on related objects.

Drawings and 3D visualisations can be automatically generated from the building model, as can specifications, quantities, ordering and tracking information and information relating to post-occupancy management. The BIM approach has a wide diversity of possible applications; for example, it can be used for thermal performance management, checking the safety of designs, site selection and fire response management processes.

Recent research has started to explore other design dimensions: moving from 3D to nD modelling which aims to integrate an nth number of design dimensions into a holistic model which would enable users to portray and visually project the building design over its complete life cyle.

Where information management systems are accessed by users in different physical locations and/or from different organisations project extranets as online collaboration platforms are used. These provide a secure, centralised repository which allow users to have controlled access appropriate to their particular project roles and responsibilities.

Summary

Successful information management requires a harmonious blend of both social and technical (supported by ICT) interactions. The ability to share and reuse information is pertinent to adding value to information management. This entails the controlled application of agreed protocols, so the effective use of information is as much about collaborative practices as it is about technology. The success of any information management system is highly dependent on the type of interactions between the participants of the project and the way it is used.

12 Value management

Value can be quantified by the following equation:

$$Value = \frac{Function}{Cost}$$

Value management is about ensuring that a project attains the optimal functional cost balance, with the proviso that the function is not reduced below a specific level. It consists of value analysis to determine what value means in meeting a perceived need by clearly defining and agreeing project objectives, and value engineering which is the process of determining how the values are best achieved.

Cost does not necessarily mean the pure capital cost of procuring and installing equipment, but will cover other value criteria. These are other associated costs – operational and maintenance costs, cost of finance, disposal costs, costs associated with design – plus non-financial costs – environmental impact, maintainability and reliability, life expectancy and construction programme – all of which needs to be considered.

Building services engineers should always design with value in mind, but due to the fragmented nature and complex interactions of the design process, value may get lost, so it may be prudent to integrate a dedicated value management activity into the project management process. By going through the value analysis stage with clients it also gives a further opportunity to gain a better understanding of the client's requirements and for clients to reflect on their prime requirements. The value management process may also encourage innovation by providing a forum for creative and lateral thinking.

The value management process may be facilitated from within the design team or make use of a third party facilitator. Whichever way is used, the process usually involves workshops which include brain-storming to encourage creative and lateral thinking, implementation of a decision making methodology and a method of reporting the decisions made. Attendance and participation by all relevant parties is vital and, in addition, participation from professionals not directly involved in a particular project can add a fresh perspective – especially if they have 'lessons learnt' to share.

This process will identify a schedule of possibilities. These can generally be categorised as alternative procurement routes for all parts of the process, materials – especially those that are mainly for aesthetic reasons – and scope reduction.

Particular technical areas for focus for building services engineers are:

■ utility services sizes
■ design criteria
■ buildability
■ specification of materials.

Cross-discipline thinking is also important, meaning that all or part of a design solution may be transferred to another discipline.

Utility services sizing

Estimates for the utility services demand will affect the cost in terms of:

■ capital cost of procuring and installing equipment
■ cost of providing space for the equipment
■ utility service providers connection costs
■ ongoing costs associated with availability of capacity.

Hence, it is desirable to ensure an accurate estimate of the capacity of the service required. In addition, a spare capacity (often 25%) is arbitrarily added. In terms of value analysis the following should be challenged. Spare capacity is required to be available at day one for future anticipated requirements. This may be because of:

■ physical expansion of a building – an extension or additional floors or basements
■ change of use – with the new use requiring an additional load.

Future changes in technology may be cited as reasons, but these should be supported by evidence.

Design criteria

Clients should be encouraged to review those design criteria that are more stringent than statutory requirements. It is important for clients to understand the implications of their requirements. It may be acceptable,

---

Changes of use

A school building may appear to be designed for a very defined use on day one. However, government policies on curriculum may change, necessitating an area previously used as an ordinary classroom to be converted to a computer suite or cookery demonstration area, both with significantly extra demands for electricity, water and possibly gas. However, it is unlikely that **all** classrooms would require such conversion and it may be prudent to only apply a spare capacity for utility services to a certain number of classrooms.

---

some of the time, for the indoor design criteria to be out of range, or for buildings to be zoned into larger areas.

The design criteria for buildings may include requirements for future flexibility of the building, for example to allow floor plates to be partitioned in very many ways. This will introduce costs.

This may include challenging existing standards and 'best practice', as those in use may be out of date.

Buildability

This involves adding value to the design process by using the experience and knowledge of construction professionals in terms of construction logistics, cost, programme and site waste.

Construction logistics determine how materials are handled from the point of origin to the point of installation, such as optimising component sizes for handling, lifting and placing in position. It also includes consideration of laydown and storage areas and installation sequencing.

Cross-discipline thinking

Building services engineers may need to undertake some lateral thinking to ascertain whether value might be added if part of their design is transferred to other disciplines, or to transfer some issues over to the building operators and end-users.

Trees

A heavily planted area will require a demanding irrigation system, comprising water supply, water storage, distribution system and all the associated pumps and controls: all requiring maintenance and on-going costs associated with the water supply.

Value analysis may determine that a plethora of vegetation, flora and fauna is an absolute requirement or that it can be reduced. The value engineering process may find an acceptable way of using artificial vegetation, either totally or integrated with the real stuff which would reduce the costs but with no detriment to the value.

Access to ceiling void services

In an ideal world all building services in ceiling voids will be fully accessible via demountable ceilings and a fully tiled ceiling be provided. An alternative would be that a (cheaper) plastered ceiling is provided with access panels at strategic locations or that there are no access panels but if access is required the ceiling is cut out as and when required and replaced and made good afterwards.

Summary

Value management recognises, measures and improves value delivered by a project. To be successful it involves thinking 'outside the box' and being prepared to challenge design norms.

13 Planning management

Planning management of projects is concerned with ensuring that activities are completed to achieve project objectives. This involves defining the tasks, estimating the required resources and durations for individual tasks and identifying the interactions between the different tasks and plotting their execution against predetermined milestones (either cost or time). This is used to monitor and control the progress of the project. The term 'programme management' is used interchangeably with 'planning'.

Building services engineers interface with programme management in two ways, firstly, associated with respect to their own duties associated in achieving the required timings of information release to the construction team, and secondly, with respect to the contribution of their deliverables to the construction process. There will be a programme management function associated with the design team, and another counterpart function associated with the construction team.

During the preparation, design and pre-construction phases, a programme will be developed to plan the delivery of the design information. This may be owned and led by clients, architects, quantity surveyors or project managers. If the project scope is predominately building services engineering, the programme may be led and managed by building services engineers.

Once the construction teams are appointed, they will develop the main programme for the works on and off site, which will include subcontracted activities, material deliveries and relevant activities in order to deliver the project to the satisfaction of the client. The programming of building services engineering site works needs to ensure that the construction team is given clear possession of large working areas.

Building Services Design Management, First Edition. Jackie Portman.
© 2014 John Wiley & Sons, Ltd. Published 2014 by John Wiley & Sons, Ltd.

In practice, it is often more efficient to delay bringing building services engineering contractors on site until unobstructed access can be provided. Delaying the start date for the building services engineering installation will not affect the overall project duration as long as unobstructed access to working areas can be provided.

During the construction phase the contractor's progress is monitored against the plan and regular progress reports are provided.

Issuing design information on the basis of predetermined schedules programmed to achieve the required timings of information release to others does not consider the internal logic of the design process – such poor planning practice is a factor in poor information management. If information transfer is not properly managed parties will not have the right information at the right time or may be overloaded with unnecessary information. This creates the risk of failure of design tasks, deficient analysis and erroneous decisions, with potential for waste in the process due to rework. Furthermore, the erratic delivery of information and unpredictable completion of prerequisite work can quickly result in the abandonment of design planning thus perpetuating a cycle likely to create further difficulties.

Design process models can give a reasonable explanation of real design processes by reflecting the methodology and characteristics of the design process and proposing strategies about how to proceed in the design process. Various entities, including professional institutions, governments and clients, have produced maps, models and procedures in an attempt to generate standard design procedures and processes which recognise the idiosyncrasies and requirements of the design process. A lack of commonality in the contemporary understanding of the design process means that a number of models of design have been developed. Some of these are rather theoretical and generic in nature.

The preparation phase is challenging to programme, as clients' requirements are still evolving and fundamental decisions still have to be made. These may rely of hard-to-get information making it difficult to formalise the time element of programmes.

The design phase was historically manageable with simple planning and management techniques based on the production of deliverables such as drawings and specifications and articulated using bar charts and drop lines. Planning methods often used in construction, such as the critical path method (CPM) and programme evaluation review technique (PERT), are highly sequential in their view and cannot model the iterative details of design processes. The format adopted is normally that of a Gantt chart. Each activity is listed and given a timescale. The activities are related to each other to build up a full profile of the project. However, these do not properly account for the creative and iterative nature of the design process and it is more appropriate to control the design process based on the production of design information rather than deliverables. Also, it has become more complex as a result of factors

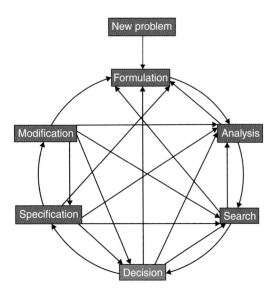

Figure 13.1 An example of a typical model of the concept design process (adapted from Jensen and Tonies, 1979).

such as fast-tracking and the increasing complexity of the fabric and content of buildings requiring greater coordination effort.

An understanding of information flows and dependencies both within discipline and between disciplines to understand how their work contributes to the whole building design process is essential. If the identification and coordination of cross-disciplinary information, essential for a fully integrated design, does not involve the design team members themselves then a a poor quality design programme results with implications for the coordination of design disciplines and general process control.

There is an intense flow of information between building services engineers and the other stakeholders to a project. Process modelling tools in the architectural and engineering design industry need to take cognisance of the highly interactive nature of the process. These iterations of redesign will refine the functional requirements, design concepts and financial constraints. Figure 13.1 illustrates the principles.

Despite that prevailing paradigm regarding multidisciplinary working, in reality the architect and the structural engineers still tended to lead the process of planning the building and so were in a position to provide information first. Thus, the building services engineers' information would always appear later. This is amplified with respect to dealings with utility service providers and specialist subcontractors, where building services engineers also had to rely on information from a large number of external parties with whom they had often no contractual linkage.

There are contractual implications of a programme's creation, approval, acceptance, management, modification and updating. Contract schedule clauses, when they exist, provide a starting point, declaring the purpose of the programme and how it is to be used and maintained on any project, and its use as a tool for determining time-related changes or impacts. Even if information is 'issued', it can be difficult and subjective to measure its completeness due to the required coordination with information produced by others and all regulatory requirements. Over the years, a body of case law has evolved, giving shape to general requirements for the creation and subsequent role of programme in proving or defending claims for delays, acceleration, impact and lost productivity.

Summary

Planning management is supported by tools and processes which need to be conducive to the iterative and multidisciplinary nature of the design process. These need to address the iterative and poorly defined nature of the design process, the lack of a method for measuring the volume, rate and effectiveness of information flow and a number of subjective elements. Also, design solutions are not absolute answers, but rather are the product of negotiation, agreement, degrees of compromise and satisfying the design criteria, thus any assessment of success must be relatively subjective.

Reference

Jensen, R. W. and Tonies, C. C. (1979) *Software Engineering*. Englewood Cliffs, NJ, USA: Prentice Hall.

14 Commercial management

Commercial management of projects is concerned with the contractual and cost issues relating to projects, from project inception to handover and beyond. Building services engineers will have to deal with their own internal commercial issues, such as fee invoicing and cash flow. This book covers commercial management issues at a project level. This involves determining procurement routes, cost and expenditure management, and dealing with contract variations, claims and disputes. There will be a commercial management cohort associated the design team, and another counterpart team associated with the construction team. The term commercial management is used interchangeably with 'quantity surveying'.

Procurement routes

The commercial management strategy will determine how building services engineers are appointed and how their duties are carried out. There are a plethora of procurement options, and the selection process needs to be carried out in a disciplined and objective manner within the framework of the client's overall strategic project objectives in order to gain best value from the marketplace. The project strategy needs to be

Building Services Design Management, First Edition. Jackie Portman.
© 2014 John Wiley & Sons, Ltd. Published 2014 by John Wiley & Sons, Ltd.

defined before the procurement route is chosen. Only with the project strategy in place can the appropriate procurement route then be chosen. There is no 'sure-fire' formula to guarantee which of the range of options will deliver the best performance for any particular project: no mutually exclusive sets of criteria uniquely and completely determine the appropriate procurement method for a specific project. The final choice will depend on the criteria profile of the client's requirements and will be influenced by the client's experience, needs and attitude to involvement in the process. More specifically, this will involve:

- familiarity with any particular procurement option
- programme constraints such as key dates and their criticality – some projects are dependent on project completion dates, or in more complex situations, windows of opportunity for – for example, schools may need to be completed before the start of a particular term – not achieving this may have a knock-on effect with consequential cost issues. Some projects, such as the Olympic Games have totally immovable dates
- cost certainty – price and the stipulated time and knowledge of how much the client has to pay at each period during the construction phase – clients may dictate certain requirements of the procurement route, for example to ensure cost certainty
- flexibility in accommodating design changes, especially if the client's requirements are subject to further development
- attitude to risk and the stance on assigning risk to the party most capable of dealing with it at the minimum cost; for example, risk due to unknown details of the utility services in the ground could be retained by the client, and managed by ensuring that site investigations are completed prior to contract negotiations or identified as potential risks within the risk management strategy and contractors cost the risk
- ownership of the finished building and its operation – for example, under PFI procurement clients forgo ownership and operation of their buildings but do have the capital asset value of the buildings
- attitude to disputes and arbitration.

Cost management

Ideally, the cost of a project would be known before commitment is made to embark on the works; this would be a fixed sum that would not change throughout the course of the procurement. Building services engineering installation on many projects can account for 40–60 per cent of the total capital cost of construction, and maybe more in particular refurbishment projects. Cost variance is a technique used to measure design performance. Cost is not just confined to the tender sum, but is the overall cost that a project incurs from inception to completion.

Typical cost elements to be considered in building services engineering life cycle costs are:

- capital costs:
  - client's costs
  - land acquisition fees (including surveys, searches and legal fees)
  - building services engineers and other design team members' fees
  - specialist consultants' fees
  - construction costs – including costs arising from variations, modification during construction period and the cost arising from legal claims, such as litigation and arbitration
  - FF&E equipment
  - legal fees
  - utility services connection costs
- finance costs:
  - cost of borrowing money for funding project
  - taxation
- operational costs:
  - Energy costs
  - Costs for moving staff and FF&E equipment into the new building.
- maintenance costs:
  - staffing and resources
  - planned routine maintenance
  - corrective maintenance
  - plant replacement and system upgrade
- residual costs:
  - removal and disposal
  - salvage value.

As part of the business case analysis clients will have undertaken a budget study to determine the total costs and returns expected from the project. This forms the start of the cost plan. This may include contingencies. Next, as more information starts to become available, with input from the various design team disciplines, the cost plan develops and becomes more detailed and accurate. The process begins with budget estimates.

Budget estimates for building services engineering services may initially be based solely on the gross floor area of a building, but using gross floor area as the sole descriptor for determining building services engineering out-turn costs is not necessarily reliable because the costs do not have precise linear relationships with building form and building function. These are gradually refined to a more detailed cost estimate as more information is produced.

If specialist systems (e.g. pneumatic tube systems) or equipment or materials that are new to the market, quotations from suppliers may be included in the cost plan.

Adjustment should be made to the estimates to account for market conditions, local variations in labour costs and, if applicable, the costs of plant and materials. These adjustments, together with adjustments for overall economic issues may be calculated from published price indices. Allowances for contingencies are greatest at the start of the project when there are the greatest number of possible risks, but can then be reduced as more details are developed and decisions are made, and some risks are passed or overcome.

Eventually actual prices are provided by the construction team, either as a tender sum or as a negotiated sum. This provides a starting point for the contract sum, the price agreed with a contractor and entered into the contract. At the end of the project the final account is agreed. Differences between the contract sum and the final account may be due to variations (with cost implications), fluctuations (in-line with pre-agreed indices), provisional sums (allowances which may or may not be expended), fees for statutory approvals and loss and expense due to client-related issues.

Due to the shorter life cycle of building services engineering services compared with 'bricks and sticks' more life cycle costing considerations are needed. This is particularly significant in PFI type projects where it is a significant consideration in the risk allocation.

Design creep

Design creep happens when there are small changes to the original brief which accumulate. These might be a change of specification for light switches, adding a 'few more' drinking water points or having a slightly better looking radiator in the entrance lobby. Added up, these small changes can have a significant effect on costs, which need to be monitored as they go along, so that clients are aware of the implications and have a chance to curtail them – unless they are happy to bear the additional costs and any programme implications.

Bills of quantities

Bill of quantities are project specific measured quantities of the items of work and materials shown in the completed design information, that is drawings and specifications. The quantities may be measured in unit items or per length, area, volume, weight or time. However, buildings services engineering designs are not usually measured in the same amount of detail. Barriers preventing quantity surveyors from measuring building services engineering include inadequate knowledge of building services technology, traditional practices which regarded building services engineering as a specialist technical area and therefore not often measured, non-completion of building services engineering services design before tender, non-involvement of specialist contractors during design and late involvement of building services engineering organisations.

Contract variations, claims and disputes

Contract variations are changes in the nature of the works described in the contract documents. They may be due to a design change, a change of material or equipment supplier or a change in the requirements of a utility service connection. Most contracts will contain provision for variation to be issued as instructions. These will either impose an increase, decrease or produce no change in the contract cost, and/or extension, reduction or no change in the programme.

When judging the impact of a variation, the costs are not simply associated with the costs of the materials and installation. They should include consideration of the necessary coordination of work between all the relevant members of the design and construction team especially if the works have started on site.

Claims may arise from building services engineers who believe that they are being required to undertake work in addition to the scope of work in their appointments.

Claims from the construction team for extra money and/or extension of time occur for very many reasons, including those resulting from building services engineers issuing information to the construction team after construction has begun. Despite using the best building services engineering entities and quality management systems, the thought of design information having no omissions or unclear specifications is unrealistic. Such new information will change the baseline information allowed in the contract sum and/or programme. Alternatively disputes can arise due to arguable deficiencies in the baseline information that do not become apparent until the construction phase starts, or maybe not until the commissioning stage.

Summary

Commercial management is concerned with ensuring that all costs associated with projects are managed throughout.

15 Quality management

Quality management of projects is concerned with preventing quality problems through planned and systematic activities from project inception to handover and beyond. This involves setting a quality policy, to set out the intentions, aims and directions, implementing it by means of quality planning, and monitoring it by a quality control regime.

The quality management process should improve the feedback cycle to create a self-improving quality and it can increase efficiencies and eliminate unnecessary costs from errors and mistakes. Quality assurance is an administrative system concerned primarily with clarity of information, which aims to prevent non-conformance by the use of clearly stipulated and authorised administrative systems and audits.

The problems of defective designs are complex and deep rooted, influenced by many factors operating at individual designer, company and construction industry, and at global or national levels. Previous and current research has considered solutions based on using ICT based systems, such as artificial intelligence, or procedural remedies such as quality assurance systems, processes, regulations and legal obligations. However, these do not necessarily address the mechanism by which the errors first originate. Despite all the valuable technological drivers to allow multi-location working, research has found that the close proximity of teams, or better still collocation, is a contributory factor in improving measured performance.

Fundamentally the QA process audits the checking process to ensure that a process has been put in place, with appropriately qualified and experienced staff, checking the required deliverables and how they were developed. The quality assurance process should not challenge the

Building Services Design Management, First Edition. Jackie Portman.
© 2014 John Wiley & Sons, Ltd. Published 2014 by John Wiley & Sons, Ltd.

design solutions, but should be checking that appropriate due diligence and process has been applied to get to that solution.

It is usual for building services engineering deliverables to be reviewed at the completion of each stage, although some projects warrant interim reviews. These reviews should embrace:

- checking that the extent and scope of deliverables matches the requirements in the appointment
- checking the integrity of information provided and relied upon provided by others; for example, are drawing backgrounds the latest revision, and how has third party information been validated?
- confirming which legislation, standards and codes are being followed – with any derogations or deviations accounted for
- demonstrating that necessary coordination has taken place with the other members of the design team (including the costing functions)
- ensuring that any required third party approvals have been processed

With respect to calculations, there are numerous 'free software downloads' available over the internet and from suppliers. Where software packages are used, assurance is required that they have been validated and are suitable for the intended applications. This should be shown in the QA/QC plan.

Summary

Defective designs have an adverse impact on project performance and on the participants and are responsible for many of the construction failures. Design quality management is hard to quantify as it consists of both objective and subjective components. While some indicators of design can be measured objectively, others result in intangible assets, depending in part on the subjective views, experiences and preferences of those asked.

Quality should already be in-built in the 'design-persona' of building services engineers. As such a quality assurance system should not require any great change in their ways of working. Much of what is required should already be done; it merely needs the formalising, recording and executing of quality procedures.

16 Performance management

Performance measurement and management provides a means of distinguishing between perception and fact which, in this context, is for the purpose of improving design process performance of building services engineers. Measurement and analysis of performance indicators can help managers to learn and make more effective decisions. The principles are illustrated in Figure 16.1.

Historically, there were clearer and more realisable benefits resulting from improving performance on site, both in terms of management and construction techniques, and therefore much industry research has been dedicated to this area. The iterative and poorly defined nature of the design process, the lack of a method for measuring the volume, rate and effectiveness of information flow and a number of subjective elements makes it more challenging. Also, design solutions are not absolute answers, but rather are the product of negotiation, agreement, degrees of compromise and satisfying the design criteria, thus any assessment of success must be relatively subjective.

Reasons for undertaking performance measurement include:

- quantifying both the efficiency and effectiveness of actions
- improving design process performance
- indicating the status and direction of a project
- providing a basis for selection of team members.

Performance measurement can be divided into three levels – individual, project and organisational. Although the performance of individuals

---

Building Services Design Management, First Edition. Jackie Portman.
© 2014 John Wiley & Sons, Ltd. Published 2014 by John Wiley & Sons, Ltd.

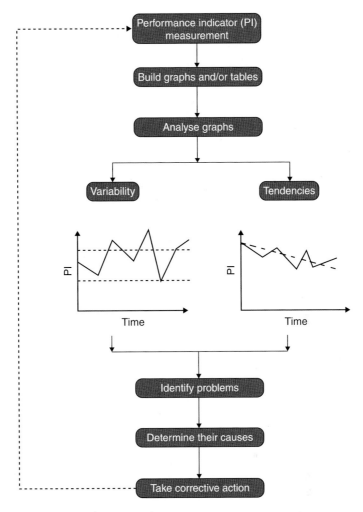

Figure 16.1 How performance indicators support management actions.

matters as they contribute to projects, this book considers performance management at the project level.

Success criteria are the measures by which success and failure of a project will be judged. Despite the widespread attention paid to the topic, there are as yet no universally accepted frameworks for assessing project success. Different people, however, assess the success of projects in different ways, and at different times, and each industry, project team or individual has its own definition of success, with project success criteria varying from project to project.

In approaching issues of design quality, a number of general features of design need to be embraced; for example, that good design often results from complex and uncertain starting points, that the process is often evolutionary and non-linear, involving interdisciplinary approaches and that it results from iterative cycles of cumulative development, where satisficing decisions are acceptable, rather than optimal results.

Traditionally, the focus has been on the three success criteria of cost, time and quality. Consequently, performance measurement in construction projects has been dominated by the conventional measures of time, cost and quality – the 'iron triangle', which focuses on short-term aspects of performance crucial for clients' immediate project success. This is now generally regarded as being rather crude and providing overly simplistic measures of performance – traditional drivers must be replaced with other measures of success.

Design is characterised by poor communication, lack of adequate documentation, deficient or missing input information, poor information management, unbalanced resource allocation, lack of coordination between disciplines and erratic decision making. Thus, developing performance indicators that can effectively and objectively account for the complex intrinsic nature of the design process is potentially very difficult. Measuring and valuing the quality of design has therefore become a key issue confronting designers and construction practitioners in order to provide competitive advantage by addressing the ever-rising client requirements and expectations for improvements in the cost, timing and quality of construction output.

The major frameworks of performance measurement in the construction industry are the European Foundation for Quality Management (EFQM) excellence model, the balanced scorecard model and key performance indicator (KPI) models. KPIs provide a tool to benchmark performance both at strategic and at an operational level, such as rectifying defects and meeting the expectations of clients with the aim of motivating participants by establishing realistic goals demonstrated to be achievable. However, KPIs lack a holistic viewpoint on the relationship between the different indicators.

Issues with performance measurement systems

Performance measurement systems may not be effective in addressing radical design performance or major improvements in the design process. The meaning of the term 'design quality' is itself poorly defined, yet ambiguity may serve a purpose. There is a need for openness, rather than limiting definitions of applicability solely to routine design. Design is about speculation, invention and innovation, with ingenuity in problem solving. Some of the features that might be measured in order to improve construction processes might relate closely to those that provide additional value during design. Some performance measurement systems can negatively influence patterns of behaviours in the organisations they monitor, leading to conservative, non-risk-taking behaviour and manipulation of statistics. Overly bureaucratic performance measurement methods can become an 'end in themselves' and lose focus on the benefits.

A major outstanding question for design performance measurement is whether improvements in design alone can lead to step changes in performance. A changes in processes can only be effective if they are tied to an optimisation of the whole production process. This implies that improvement in design needs to be part of wider performance improvement programmes to be most effective.

Typically, if there is a mismatch between performance measures and the real practices and needs of designers, individuals will object to what they see as time-wasting 'number crunching exercises'. Design performance measurement works best when design (and related) staff are involved in data collection and interpretation, whereas KPIs that focus on benchmarking projects retrospectively are of little use for control during the design process, as they do not provide the opportunity to change.

Summary

There are many different proposed systems for the measurement of performance within the construction and related industries, yet there is no general consensus as to whether or not these systems are based on appropriate metrics or are capable of being implemented on a broad scale Although the design process has only recently been considered with respect to performance measurement, and its monitoring is very challenging, it is generally acknowledged that it is important to do so. However, due to the ambiguous definition of project success and the different perceptions of participants toward this concept, it may be difficult to tell whether a project is successful because there is a lack of consensus.

Part Four Special buildings

17 Special buildings

There are certain buildings that require special attention with respect to the design of the building services engineering. This chapter does not prescribe details of design, but rather highlights the particular design management issues and how they might be approached. These might be associated with having to deal with particular stakeholders or regulatory conditions, requirements to include special building services engineering systems or for specific variants on regular building services engineering systems solutions.

17.1 Commercial kitchens

The main purpose of commercial kitchens is to deliver menus on a large scale. The process involves deliveries of incoming food and materials, appropriate storage, preparation and cooking of dishes, passing over for service, receiving payment, receiving dirty dishes and washing up, and recycling and disposal of food and other waste. Supporting facilities may include offices, staff restrooms, toilets and showers and cleaners' rooms.

Building Services Design Management, First Edition. Jackie Portman.
© 2014 John Wiley & Sons, Ltd. Published 2014 by John Wiley & Sons, Ltd.

The requirements for the building services engineering systems are driven by the menus, the times of the meal services and the strategy for delivery. Provision for vending machines may also be included. These requirements can only be determined by clients on behalf of end-users. Often a catering consultant is involved who will work with clients to develop comprehensive catering concepts, menus, kitchen planning and adjacencies work for throughput, specification of kitchen equipment and payment mechanisms.

The strategy for service delivery has the biggest influence on the kitchen design. The main options are:

- *cook and serve* – meals are served to diners immediately following preparation. The main challenge is the maintenance of service temperature during the holding time before service
- *cook and hold* – the conventional cooking process in catering services where meals are prepared in the kitchen then kept in the appropriate conditions either for delivery to diners on demand for holding in a buffet presentation
- *cook and choice* – a food production process during which the diner has an active influence on the preparation. The diner can, for example, determine the choice of ingredients and their relative proportions
- *cook and look* where meals are prepared outside the kitchen, directly in front of the diner, who is able to watch as the meal is prepared. Show cooking provides variety and entertainment.

A 'regen' philosophy may be used in connection with any of these strategies. Following cooking, foods are either subjected to shock freezing at approx. −40 °C and then stored at −18 °C, or shock chilling within 90 minutes to 1–3 °C. These may be prepared remotely from the place of service, transported to the kitchen and, shortly before consumption, the frozen food is thawed and reheated to complete the cooking process.

Key issues for building services engineering design are associated with ensuring appropriate indoor environmental conditions for safety and for hygiene of foods.

Temperature control

Appropriate temperature conditions are required for the processing and storage of products. This may be vegetable stores, wine stores and freezer stores. It may be prudent to provide a temperature alarm linked to such rooms.

Ventilation

Catering and cooking can produce significant amounts of steam, odours, fumes, grease and vapours, as well as large amounts of heat, so the design of the ventilation is significant. Poor ventilation can lead to an uncomfortable working environment and incomplete combustion at fired appliances leading to the risk of carbon monoxide accumulating in the space. The design should be arranged to be easy to clean, avoiding build-up of fat residues and blocked air inlets, which lead to loss of efficiency and increased risk of fire.

Cooker hoods are used to remove cooking fumes at source. They have several elements: wall mounted or island, box form or slanted, integral or external fans, the provision of lighting and the introduction of make-up air via the front or perimeter of the hood, cooker fire suppression system and a link to the fire alarm system.

Adequate ventilation and extraction should be provided to ensure that steam emitted from the dishwasher does not give rise to condensation within the room. Consideration must be given to the installation of a heat recovery system within the dishwasher to reduce the requirement for a dedicated extraction system to the dishwasher and provide an energy-efficient system.

Fresh air can be introduced either via mechanical or natural means (or a combination of both). This air may require to be tempered, and control of pest entry must also be considered.

The pressure regime is important to combat the release of undesirable odours, smoke and steam into other spaces to mitigate the potential for cross-contamination which this could cause. This may be achieved by designing for positive airflow between critical areas.

Disposal of kitchen waste

Kitchen waste contains significant levels of fat, oil and grease. These can cause major problems to drains and sewers if their disposal is not properly managed, such as blockages when the fat, oil and grease in liquid form cools, congeals and hardens, and sticks to the inner lining of drainage pipes restricting the wastewater flow causing the pipes to block, or when they enter rainwater pipes or gullies they can cause pollution in streams and rivers.

Grease traps remove fat, oil and grease from wastewater before it reaches the public systems. They are specially designed units which are placed in drainpipes to separate the fat, oil and grease from the rest of the wastewater. The wastewater then continues to flows through the public system to the treatment works while the grease is retained in the trap to be collected by a licensed waste oil collector at regular intervals. Without a grease trap, obstruction and smell may be caused by grease build-up.

Food macerators are used to chop and grind food into small pieces prior to disposal into the drainage system. However, because fat combines again in the pipes, resulting in blockages, an alternative strategy is placing waste food in bins. Recycled waste cooking oil can be collected either to be used for bio-diesel for transport fuel or for incineration for the generation of electricity.

Adequate floor drainage gullies should be installed to allow direct discharge from defined items of catering equipment and to allow appropriate drainage to assist floor cleaning procedures.

Fire engineering

The likelihood of a fire in kitchens is high due to the increased number of potential sources (bare flames and equipment with high surface temperatures), exacerbated by the proximity of fuels (natural gas, grease, oil and fat) which support the growth and spread of fires. Thus it is necessary to take extra precautions to prevent a fire starting in the first place, extra vigilance on detection and, if a fire occurs, the means to contain it as quickly as possible.

Provision of a gas interlock with the ventilation stops the gas supply being turned on until the ventilation system is operating correctly, thus preventing the build-up of dangerous gases such as carbon monoxide and nitrogen dioxide. The gas interlock should also be interfaced with the fire alarm system, so that the gas can be switched off during a fire situation.

Fire detectors should be provided to cover the entire floor area, and it may also be necessary to locate them within ductwork and ceiling voids, so as to provide the earliest opportunity for detection. Using traditional smoke detectors for kitchen areas initiates false alarms due to the smoke from cooking or vapour from hot water; thus either smoke detectors with sensitivity settings or heat detectors should be used to avoid such instances.

If kitchen canopy extract hoods are installed above cooking areas, they should incorporate smoke detectors linked to the gas solenoid, fire shutter and fire alarm, so that in the event of extract fan failure, fire in another area of the building or kitchen, all systems shut down and the kitchen will effectively become a fire compartment to limit spread.

Electrical services

The design of the electrical services needs to recognise that the electrical demand in the kitchen is a function of how it operates, the operational areas of kitchens include wet and hot areas, and there may be a significant amount of exposed metal.

There is often significantly more equipment in a kitchen than is ever used at one time. The sequence is driven by the menu, which needs to be understood when developing the electrical load profile. Typically the final selection of kitchen equipment may not happen during the design stage. With respect to the utility services provisions (electricity, gas, water and ICT connections), the particular requirements for any type of equipment – such as an oven or fridge – may vary greatly from one manufacturer to another, so it may be prudent to add a note during the design stage advising that the equipment's utility load and supply arrangement are tentative, for design purposes only, and that during construction these will need to be reviewed.

There are other points that need to be considered:

- For the purposes of electrical safety, all major kitchen equipment should have an emergency stop push button to disconnect the power, or one provided at central points.
- Switches for equipment should be located to avoid damage from heat.
- Socket outlets must to be located away from any water tap and not where they might be liable to catch fire due to heat transmission.
- It may be prudent to label plugs, so that they are not accidently swtiched off.

In addition to the electrical loads that require emergency power for life safety reasons, there may be requirements for standby power for operational reasons for:

- cold stores
- freezer stores
- live fish tanks
- live cooking stations.

These need to be discussed and agreed with the clients: building services engineers cannot be left guessing.

Lighting system

Luminaires need to be installed in such a way that will not contribute to food contamination. They should also be designed and installed in a way that facilitates ease of cleaning. All luminaires are to be minimum IP54 and preferably connected to the same phase.

The provision of emergency standby lighting should be considered, in addition to escape lighting. If the mains power fails, it may be prudent for staff, and maybe even diners, to remain in the spaces.

Maintenance considerations

The design of the building services engineering systems needs to mitigate the accumulation of dirt, contact with toxic materials or the shedding of particles into food. The systems need to be easy to clean; for example, integrated ceilings might be provided which contain integral, rather than suspended, lighting and ventilation systems in a sealed unit with a hygienic surface that is easily cleaned and maintained.

Anti-microbial finishes, which prevent significant amounts of harmful bacteria growth, may be considered for surfaces such as light switch plates and wall-mounted trunking.

17.2 Hospitals and healthcare facilities

Hospitals are multifunctional buildings that may comprise a wide range of facilities which can broadly be classed as:

- outpatients facilities
- diagnostic and treatment facilities
- in-patients facilities
- hospitality facilities (food service, laundry, cleaning, waste management and general housekeeping)
- research and learning facilities
- administrative facilities, to support functions (patient admissions, documentation and records and finance)
- secondary areas, such as retail areas, food courts, public spaces, exhibition spaces, childcare areas and conference facilities.

Clients may enlist specialist healthcare planners to assist with the design of a building. They have specialist knowledge of the healthcare processes in a building, how these are driven by clinical policies and strategies.

The business case analysis is very complicated due to the diversity of technical requirements for service delivery and clinical strategy, and the large number of stakeholders. These include hospital staff, patients and visitors, both present and future. There may also be significant public involvement with inputs required from the likes of Members of Parliament, education providers, research establishments, charities and regional agencies. Drivers include the need to optimise patient safety, quality and satisfaction, as well as other building users' safety and satisfaction, all constrained by a financial plan that includes energy efficiency.

Clients may enlist specialist healthcare planners to assist, having knowledge of the service's clinical strategies.

This diversity is reflected in the breadth and specificity of regulations, codes and practices relevant to the design of hospital and healthcare facilities. Each of the wide-ranging and constantly evolving functions of a hospital include highly complicated building services engineering input. The building services engineering design, and hence the design management, is consequently complex. The capital value of the installed building services engineering systems in hospitals is one of highest for any type of building, so the running costs are also significant.

Key issues for building services engineering design are associated with maintaining the continuous operation of the facility and incorporating significant amounts of FF&E. This means that solutions will need to consider cleanliness and sanitation in the short term, and flexibility and expandability in the longer term.

FF&E fit-out in hospitals includes significant equipment, such as MRI scanners, which have specific requirements for power supplies and removal of heat gains.

Infection control has a significant impact on cleanliness and sanitation, with ventilation systems having a significant role in reducing the risk of infection due to air-borne transmission. Special materials, finishes and details are used for spaces that have to be kept sterile, such as integral cove base. The new anti-microbial surfaces might be considered for appropriate locations. Incorporating O&M practices that stress indoor environmental quality (IEQ) needs to be addressed.

Since medical needs and modes of treatment will continue to change, hospitals should be able to adapt, often while still in operation, so consideration would be given to providing building services equipment in modular, easily accessed and easily modified arrangements.

Heating

The risk of burns from hot surfaces is a significant factor in the design of the heat emitters. Possible considerations are:

- radiant panels – this type of low surface temperature heat emitter uses low temperature water and a very large surface area for transferring the heat to the space – in the form of a radiant panel mounted on the wall. However, these are relatively impractical for most areas of a hospital because they need to occupy a substantial area of the wall to be effective, and are quite expensive in terms of capital costs and installation. In many older properties, the ability of the walls to support the significant weight of a radiant panel would also need to be assessed
- underfloor heating – for spaces with high ceilings, such as atria and lecture theatres in teaching hospitals, underfloor heating may be an appropriate solution, but in smaller spaces radiators are generally a more cost-effective, responsive and efficient option
- alternatively reducing flow temperatures to standard radiator systems will reduce risk but will also reduce performance to a level where the heating is ineffective, as the system will have been designed for higher flow temperatures
- heat sources at high level – Locating sources of heat at high level is another option but, as warm air rises, this tends to reduce the effectiveness of the heating, because most of the heat will be concentrated at ceiling height
- destratification fans can be used to force the warmer down into the space, but this can cause uncomfortable draughts and it increases running costs by using more electricity, and the capital costs of such an installation are also higher
- guarding hot surfaces – in any building with a conventional 'wet' heating system, with perimeter radiators served by central boiler plant, by far the most practical solution for preventing contact with

hot surfaces is to guard the heated areas with a physical barrier. Such guarding measures need to be part of the design of the system, rather than ad hoc measures that may not provide complete protection. Temporary measures such as putting furniture in front of radiators is not considered acceptable under health and safety regulations

■ low surface temperature radiators (LSTs) – the most effective way of guarding hot surfaces is to use radiators that are designed to eliminate the risk of burns by incorporating a casing that covers all hot surfaces. LSTs are designed to do just that, providing a safe, cool-touch solution. The casing of a cool-touch LST covers the pipework as well as the radiator, so that all hot surfaces are concealed, and exposed surfaces remain at a safe temperature of no more than 43 °C. Radiators in public areas should also be encased down to floor level so that young children cannot gain access to hot surfaces from below.

An alternative to hot water and electric heating is to use steam as a heating medium. It can carry considerable amounts of heat energy. Steam is distributed by pressure generated at the boiler, whereas hot water is generally pumped. The philosophy of providing a circuit containing flow and return pipes is the same as a water heating circuit, but the arrangement is different. The flow and return pipes in a hot water system, of course, contain hot water at different temperatures, whereas in a steam system the flow pipes distribute steam, while the return pipework must accommodate condensate. The process of returning this condensate is an important factor in system efficiency.

Ventilation and air-conditioning

Ventilation and air-conditioning systems are provided in healthcare facilities, in the same way as, say, an office building to provide a safe and comfortable environment for patients, staff and visitors. However, more specialised ventilation is provided in primary patient treatment areas such as operating departments, critical care areas and isolation units. These each have their own specialist requirements for temperatures, relative humidity, ventilation levels and flow patterns (to preserve a desired airflow path from a clean to a less clean area), filtration, pressures in different rooms and even zones within rooms and to remove, contain or dilute specific contaminants and fumes. For example, the ability to deliver filtered air to surgical area where there are open wounds can help to reduce the risk of infection. Specialist ventilation and air-conditioning is required in isolation units for patients who present a biological, chemical or radiation hazard to others and/or have a reduced immune system.

A further complication may be that the design conditions selected within patient areas need to satisfy the comfort requirements of staff and patients, who often have very different levels of clothing and activity.

Special requirements for the removal of anaesthetic gases which are subject to health and safety and legal requirements associated with exposure. An anaesthetic gas scavenging system collects waste gases due to leakage from anaesthetic equipment and the patient's breathing circuit, particularly during connection and disconnection, and from the interface with the patient. These are transported via transfer tubing, to receiving devices which dispose of them safely. The air movement design should ensure that this leakage is diluted and removed from the room. Anaesthetic gases are heavier than air, so placing the supply terminal at high level with an extract at low level, adjacent to the anaesthetic-gas terminal units, will ensure that staff are in a clean airflow path.

Water services

The provision of water services generally requires the same considerations as for other buildings. However, if water supplies are required for specialist systems, for example endoscope cleaning installations or dialysis units, building services engineers should consult the hospital infection control team to establish any specific water treatment requirements for the process, and also the public or private utility supplier to clarify any special precautions that may be necessary, such as backflow prevention devices.

Electrical services

The provision of electrical services generally requires the same considerations as for other buildings, but there are some special features to be considered:

The provision of standby power needs detailed consideration with significant input from clients. Within any healthcare environment, there are wide ranges of departments with complex requirements and potential risks. The risk management process will categorise each department in terms of susceptibility to risk from total (or partial) loss of electrical supply.

Uninterruptable power supplies (UPSs) are provided where the risk of even a short interruption to the power supply is unacceptable; for example, the patient environment of an operating theatre is a clinical risk.

Isolated power supplies (IPSs) are connected to critical care and life support equipment attached to patientswho are anaesthetised and thus unable to inform staff of an electric shock. They constantly monitor for short circuits or earth leakage faults on the supply side of an electrical circuit feeding a piece of medical equipment. If they detect any fault, an audio and visual alarm is raised so that the medical equipment on the affected circuit can be unplugged and replaced or reconnected to a healthy circuit.

Bed head trunking may be used to integrate several services – such as nurse call, lighting, small power, data and medical gases – into one enclosure. These may be prefabricated off site with wiring and pipework to allow for easier and quicker installation on site.

Operating lamps are provided to assist medical personnel during surgical procedures by illuminating the required area of the patient. They can be continuously adjusted by hand, and are usually suspended from the ceiling. The light source will have a very high colour rendering index, making it as near as possible to natural light – this is important as the colour of body parts and fluids is used as a basis for medical assessments. The lux level is as high as or, in some circumstances, higher than that of natural daylight. As an interruption to light would be disruptive to a surgical process, they are connected to UPSs. One or multiple lamps may be included on the same pendant arrangement. Other building services engineering systems may be incorporated into the pendant arrangement, such as small power, data, monitors, medical gases and CCTV.

Examination lights are similar to operating lightings, but are used outside an operating theatre.

Fire engineering

Fire engineering design in hospitals is particularly challenging due to the diversity of people in the buildings: patients, staff, hospital and other visitors. This is particularly true where patients are highly dependent on members of staff, for example the elderly, the mentally ill and those in intensive care units. Lack of alertness, lack of mobility and high dependency on fixed equipment have obvious implications for patient safety in the event of fire. Balanced against this, many parts of the building may be occupied 24 hours a day, meaning that it is more likely that fires will be manually detected sooner than if relying on automatic detection systems.

Visual alarm devices may be provided as an alternative to alarm sounders in areas where an audible alarm is unacceptable, such as in very high dependency patient access areas, such as operating theatres, intensive care units and special care baby units.

The main evacuation strategy is normally to move people from the area where the fire is located to safe places, such as another ward or the protected lobby. Horizontal evacuation is often the first objective so as to avoid patients having to use stairs, but if the escape route to the protected lobby is blocked, patients are evacuated via the stairwells located at the end of each corridor.

Security systems

In addition to the general safety concerns of all buildings, hospitals have several particular security concerns associated with the protection of hospital property and assets, including drugs and medicines, the protection of patients (especially if incapacitated) and staff and safe control of violent or unstable patients. They may also be vulnerable to damage from terrorism because of the proximity to high-vulnerability targets, or because they may be highly visible public buildings with an important role in the public health system. Balanced against this, hospitals need free and easy movement for staff, patients and visitors.

Hospital lifts

Hospital lifts, as well as transporting the usual people and equipment, need to accommodate patients with a wide range of special needs, trolleys, stretchers and hospital beds and any associated medical equipment, plus any attendant medical and other support staff, and all this in buildings used by the general public. Additionally, the fire engineering solution for buildings may require hospital lifts to be integral to the evacuation strategy, so move persons with limited mobility where the usual stairways would not be feasible. Thus, hospital lifts need to be durable enough to cope with intensive use while still being safe and reliable. Other particular features are:

- ride quality, a function of the acceleration and deceleration and jerk (the rate of change of acceleration and/or deceleration), and accurate levelling are needed due to the needs of patients to mitigate any additional pain or distress
- the interior design of the lift car, which needs to cope with hospital environments in terms of hygiene and infection control, as well as the ability to wash down the interior completely. Robust surface finishes and crash rails are important considerations
- the lift control systems determine how lift cars are dispatched in response to activating a lift call request. These comprise the 'hall' buttons, outside the lift car, and the buttons inside the car. These may be simple push buttons, key-controlled, use keypads or be integrated with the access control cards – or a combination. The lift control systems may need to be configured to cope with particular aspects of hospital environments, such as emergency evacuation and special bed priority service and theatre control
- there may be two separate fire-protected power supplies to serve any lift.

Specialist hospital systems

These include:

- steam systems – often required for such processes such as autoclaves (for sterilising), food preparation and laundry, and it may then subsequently be used for space heating and domestic hot water heating
- medical gas systems that provide gases for use directly by patients, as part of medical procedures, to 'power' medical equipment or to

support laboratory functions. They consist of central or localised sources, piped distribution systems and final outlets with monitoring and alarm functions

■ waste anaesthetic gas scavenging systems which collect expired and/or excess anaesthetic gas and conveys it via pipes to the exterior of the building where it can be discharged safely

■ medical vacuum systems that support vacuum equipment and evacuation procedures, usually supplied by various vacuum pump systems exhausting to the atmosphere

■ nurse call systems that allow patients, a healthcare staff member already with the patient or a visitor with a patient to alert a nurse or other healthcare staff member remotely to their need for help. These are linked to the nurses' station, and usually, a nurse or nurse assistant responds to such a call. Some systems also allow the patient to speak directly to the staff member, while others simply beep or buzz at the station, requiring a staff member to actually visit the patient's room to determine the need

■ medical AV systems to record surgical procedures – this may extend to telemedicine, which allows for healthcare delivery at the patient's location from a remote source and where doctors or other medical expertise is located

■ pneumatic tube systems that propel cylindrical containers through a network of tubes by compressed air or by partial vacuum. They are used for transporting laboratory samples, units of stored blood or patients' files between nurse stations and laboratories. They are arrange to prohibit bacteriological transmission

■ baby tagging system which comprises an electronic tag fitted to the ankle of a new-born baby, and linked with an alarm that will operate if there is any unauthorised attempt to remove a baby

■ surgeons' control panel, which provide an information dashboard for easy reference for operating theatre staff. This includes various building services engineering system information such as ventilation controls and indication, medical gas and alarm indicators, time and time-lapse, power supply alarms (for mains, UPS and IPS systems), operating lights and battery backup controls, general lighting, 'room in use' controls, laser interlocking systems and fire alarms statuses.

Hospitals are large public buildings that have a significant impact on the environment and economy of the surrounding community. They are heavy users of energy and water, and produce large amounts of waste. Because hospitals place such demands on community resources they are important candidates for sustainable design.

Noise- and vibration-generating plant should not be housed either directly above or below sensitive areas (e.g. operating or anaesthetic rooms) unless there is no alternative, in which case additional care and attention must be given to control measures.

17.3   Data centres

A data centre is a facility used to house computer systems that provide facilities for data processing, data storage and data routing. There will be supporting administrative accommodation. An entity may have its own data centre facility within a building or it could be a standalone building; alternatively, data centres are run on a commercial basis, renting out space and/or computing capacity to customers.

Clients are particularly concerned with ensuring that business and reliability objectives are met. The tier classifications represent a standard way to define the uptime of a data centre. The Uptime Institute has established four levels of fault tolerance for data centres, with Tier 1 being the lowest level and Tier 4 the highest, with complete multiple-path electrical distribution, power generation and UPS systems.

Tier classification

The Uptime Institute, Inc. is a consortium of companies devoted to maximising efficiency and uptime in data centres and IT organisations. Member companies learn from each other through sponsored meetings, tours, networks and benchmarks. Performance standards developed by the Institute are based solely on the needs and preferences of users.

The four tiers, as classified by The Uptime Institute, are:

- Tier I: composed of a single path for power and cooling distribution, without redundant components, providing 99.671% availability. This is the simplest and is typically used by small businesses.
- Tier II: composed of a single path for power and cooling distribution, with redundant components, providing 99.741% availability.
- Tier III: composed of multiple active power and cooling distribution paths, but only one path active, has redundant components and is concurrently maintainable, providing 99.982% availability.
- Tier IV: composed of multiple active power and cooling distribution paths, has redundant components, and is fault tolerant, providing 99.995% availability.

As you can see, Tier IV is the most robust and least prone to failures. It is designed to host mission critical servers and computer systems. The cost of collocating in a data centre increases as the tier increases. Tier II and III facilities generally meet the uptime requirements of most hedge funds and investment management firms.

Data centres must be designed to be able to effectively distribute a large volume of building services engineering distribution: data and electrical power cables and water pipelines. Their design should be flexible enough to adapt to the latest technological and cooling advancements. Their design includes redundant or backup power supplies, redundant data communications connections, adequate cooling capability and appropriate fire and security engineering solutions.

Air-conditioning and cooling systems

Mechanical engineering infrastructure design addresses mechanical systems involved in maintaining the interior environment of a data centre, such as heating and ventilation. The control of heat in a data centre is a key priority. Overheating leads to equipment failures and shutdowns. Poorly designed air-conditioning may be hugely inefficient and consume vast quantities of electricity. Inadequate air-conditioning will simply be unable to cope. Data centre cooling systems must be capable of efficiently cooling high density rack loads that are variable and diverse throughout the facility.

The principle of hot and cold aisles is an acceptable practice for cabinet layout within a data centre. The aim of a hot and cold aisle configuration is to maximise the flow of chilled air across and through the equipment racks. This requires chilled airflow from bottom to top and from front to back through the racks. The design uses air-conditioners, fans and raised floors as a cooling infrastructure and focuses on separation of the inlet cold air and the exhaust hot air.

In its simplest form, hot and cold aisle data centre design involves lining up cabinets of data equipment, on a raised floor, in alternating rows with cold air intakes facing one way and hot air exhausts facing the other. The rows composed of rack fronts are called cold aisles, due to the front-to-back heat dissipation of most IT equipment. Typically, cold aisles face air-conditioner output ducts. Computer room air-conditioners (CRACs) positioned around the perimeter of the room or at the end of hot aisles, push cold air under the raised floor and through the cold aisle. Perforated raised floor tiles are placed only in the cold aisles to direct chilled air to the front of racks to get sufficient air to the server intake. As the air moves through the servers, it is heated and eventually dissipated into the hot aisle. The exhaust air is then routed back to the air handlers. The rows that the heated exhausts pour into are called hot aisles. Typically, hot aisles face air-conditioning return ducts.

A physical barrier system can be used to isolate hot aisles and cold aisles from each other and prevent hot and cold air from mixing; alternatively the arrangement of fans can minimise the air mixing.

Electrical system

The electrical power plant and distribution system design is crucial to data centre reliability and operational efficiency. Key design issues are the sources of electrical supply and ensuring that the electrical distribution systems maintain continuity and quality of electrical power, and the provision of maintenance and emergency shut down switches. This involves backup power supplies such as diesel generators, uninterruptable power source (UPS) systems and power distribution units (PDUs). PDUs are a particular type of electrical switchgear which comprise the usual incomer and outgoing ways, but they may also be able to control and monitor power to specific servers, switches and other data centre devices and assist in balancing power loads.

To prevent single points of failure, all elements of the electrical systems, including backup systems, are typically fully duplicated, and critical servers are connected to both the 'A-side' and 'B-side' power feeds. This arrangement is often made to achieve N + 1 redundancy in the systems. Static switches are sometimes used to ensure instantaneous switchover from one supply to the other in the event of a power failure.

UPSs and, in many cases, the backup diesel-powered generators, should be sized to energise all computer equipment, HVAC systems and other electrical devices (such as emergency lighting and security devices) for 100 per cent of the power demand for a period of time determined in conjunction with clients after a power interruption. Also, UPSs should be sized for 'peak' load or fault overload conditions. This relates to the surge in power demand when the equipment is first energised. As a rule of thumb, size the UPS for 150% of the operating demand.

Structured cabling infrastructure

There are extensive structured cabling systems within data centre environments, including extensive horizontal cabling.

Fire engineering

As well as the usual life safety issues, economic loss from loss of function or loss of data, as well as ensuring quick operational recovery, is a critical issue in developing the fire engineering strategy for data centres. Strict management is also required to ensure that potential risks – for example, stacks of paper, packing material and other miscellaneous combustible materials in the space – are minimised.

Automatic detectors may have much higher than normal sensitivities or the ability to operate at higher temperatures or in higher air velocities (along with deeper consideration of the associated airflow patterns). This ensures warning of a fire at an early stage. This allows investigation, interruption of power and manual fire suppression using handheld fire extinguishers before the fire grows to a large size and possibly before gaseous agents and water sprinklers have to be activated.

As well as the usual potential sources of fire (electrical switchgear and fans) the abundance of large quantities of electrical power cables, especially in floor voids acting as plenums, needs to be considered. Electrical power cables in a fault situation carry more current than they are designed for. As well as the usual automatic detectors, linear heat detectors may be installed above, or even within, bundles of power cables to provide overheat warnings and hence more rapid fire detection. Data cables do not carry electrical current and thus are not a potential source of fire.

Security systems

Physical security also plays a large role with data centres. Physical access to the site is usually restricted to selected personnel, with controls including bollards and mantraps. Video camera surveillance and permanent security guards are almost always present if the data centre is large or contains sensitive information on any of the systems within. The use of fingerprint recognition man traps is starting to be commonplace.

17.4 Shopping centres

Shopping centres provide space for a number of retail outlets and other service providers under one roof, with common facilities for the likes of car parking. This requires facilities to deal with deliveries of incoming goods directly or indirectly to storage areas, recycling and disposal of waste. Supporting facilities may include offices, staff restrooms, toilets and showers and cleaners' rooms.

The terms landlord and tenant areas are often used. Clear definitions and demarcations need to be agreed to any particular project as it may affect zoning and metering of building services engineering systems. Typically landlord areas cover all common areas meant for their customers' use including parking and back-of-house areas, while tenant areas cover the retail space, plus space for storage, plant and possibly vehicle parking.

When planning a shopping centre clients may not know the final users, or not all of them, but they need to provide some information to building services engineers to provide them with a basis for design. Particular information is:

■ the 'footfall' – the estimated population of visitors (peak and average) in the public areas
■ demarcation of landlord and retail areas – for the purpose of influencing the distribution and zoning of building services engineering systems
■ guideline on the types of tenants in the retail areas, e.g. a hairdressing salon, with a high demand for water.

Often building services engineers will be responsible for landlord areas of the shopping centre. Typically the landlord areas will be fully fitted out and the tenant areas left as shells to be fitted out to suit the tenant's needs. As such guidelines will be required with respect to the constraints, such as maximum power, water and gas demands, interfaces with building fire, emergency lighting and control systems.

17.5 Sports facilities

Sports and leisure facilities that provide a venue to participate in sporting and leisure activities will have different aims and objectives depending on a variety of criteria:

- whether they are sport specific or multipurpose
- whether they are outdoor, indoor or both
- whether they are public, private or voluntary operated
- requirement for spectator venues

Sports facilities include the space where the sport is taking place, and may also include spectator areas, changing facilities and storage areas and all the administrative support areas.

The client will be particularly concerned with ensuring that the facility achieves the standards required for the level of events required: community, club, regional, national or international. This may also affect whether the events are to be televised.

Also, there may be varying and conflicting requirements of individual sports in a multi-sport facility.

Key stakeholders may include sports facilities operators, the relevant sports clubs, community associations, national and international governing bodies.

Specialist input may be required from acoustics engineers; they will need to consider the acoustic environment from the perspectives of both the players and the spectators.

Most lighting design considers the illuminances in just two dimensions – on wall and floor services. In sporting facilities the whole of the three-dimensional volume above and including the field of play should be considered. Where events are televised, or for sports which involve great use of the height above the playing area – for instance volleyball, athletics throwing events, tennis or rugby – consideration of the full volume is especially important.

In addition to emergency escape lighting (to allow for safe evacuation when required) for sports facilities, standby lighting can be further subdivided into 'standby lighting' and 'continuation of an event'. The degree of standby lighting provided will be influenced by the nature of the activities being undertaken, the duration of the activities and the level of associated risks involved. Exceptions will be the provision of alternative cover for major events, where loss of lighting would cause unacceptable cancellation.

Public address systems may have a wider remit: emergency evacuation procedures, to communicate with inspectors, to relay entertainment and to convey advertising messages.

Swimming pools

The basic idea is to pump water, via a circulation system comprising pumps, inlets, outlets and associated pipework, in a continuous cycle, from the pool via filtration and a dosing system for chemical treatment and back to the pool again. By this circulation, the water in the pool is kept relatively free of dirt, debris and microorganisms (bacteria and viruses). Heating will also be included in the process. Water lost by evaporation, backwashing the filters, swimmers carrying it out of the pool on costumes and water lost by disposal of effluent and backwash water will need to be made up.

17.6   Hotels

A hotel, as a minimum, provides overnight accommodation on a commercial basis. This requires guest bedrooms and circulation areas. If meals and other facilities are also provided for guests additional space will be required both to make the facilities available (front-of-house areas) and also to support these facilities (back-of-house areas). Front-of-house areas can be further divided into, public areas, such as lifts, foyers and corridors, and supporting facilities, such as gyms, meeting rooms, function and pre-function areas, swimming pools and restaurants. Public areas are interior areas that are standard at most hotels, whereas the extent of supporting facilities is more dependent on the type of hotel.

Different hotels will be intended to attract different clients (business, pleasure), in different types of locations (rural, suburban, city-centre, airport, roadside) aimed at different markets (resorts, casino hotels, convention hotels, conference centres) and budgets (economy, mid-range, high-end) and hence they will need different facilities and emphasis. They may be standalone or integral to mixed-use developments. The supporting facilities may be all or partly available to non-guests.

A basic measure of the size of any hotel is the number of rooms or 'keys' and the number of covers in restaurants.

Interior designers and hotels

For certain types of hotels, particularly high-end, there will a higher emphasis on interior design, and clients may employ specialist interior designers. The scope of these interior designers may extend to the design of bespoke specialist luminaires, supply and extract grilles, radiators and sanitaryware. The building services engineers will need to review all of these to ensure that technical performance requirements are achieved – otherwise they may look good but not actually do the job. Also, interior designers may need to be reminded of the fact that building services terminal equipment has to be fed by pipes and wires.

Key stakeholders in the design of hotels are:

■ the client, who is the hotel owner
■ hotel operators who are responsible for the day-to-day running of the hotel, including hotel staff. As well as providing guest accommodation, and additional supporting facilities, hotel operators will be responsible for routine maintenance and will procure other capital projects needed for the hotel, although these will typically be authorised and paid for by the hotel owner
■ hotel staff
■ 'external' hotel service providers – some of the supporting facilities may be provided by external entities who will have their own requirements in terms of space, utility services provision, metering, monitoring and billing
■ guests
■ non-guests using supporting facilities.

Key issues for building services design are understanding the specialist hotel systems:

■ room management systems
■ hotel management systems

Multifunction areas provide challenges for building services engineering systems design as they may need to accommodate the requirements for exhibitions, conferences, banquets and weddings. Particular challenges in multifunction spaces are:

■ adapting to different room sizes. These may be larger rooms with the capability for subdividing into smaller rooms on a temporary basis; thus controls for lighting, heating, ventilation and air-conditioning need to be usable in each scenario, and similarly for the fire protection and small power solutions

- different occupancy patterns
- varying requirements for audiovisual.

17.7   Educational buildings

Educational buildings provide facilities for formal learning, requiring a range of teaching accommodation and outdoor areas suitable for different age groups, subjects and needs. These are supported by administrative and operational functions.

Key issues for building services engineering design are to ensure that the systems are safe and suitable for the needs of the learners and staff, and that the building is adequately flexible for the future. Building is subject to alteration due to changes in curriculum, and social changes in the demographics of the learners, meaning that teaching spaces may be used differently or require extensions. The building services engineering systems must be designed to allow for changes in configurations to be made easily and with minimal disruption.

Increasingly, educational buildings are used for other purposes outside the core teaching times, so the stakeholders may include more than those involved with the school itself.

Heating systems

Educational buildings may use a conventional 'wet' heating system, with radiators located around most rooms in the building, served by central boiler plant. However, standard radiators and associated pipework can reach a temperature as high as 75 °C, hot enough to cause serious burns within seconds. For this reason, it is important that younger learners cannot come into direct contact with these hot surfaces. The most practical way to achieve this is to use radiators that incorporate a casing that covers all hot surfaces. Low surface temperature (LST) radiators are designed to do just that, providing a safe, cool-touch solution. The casing of a cool-touch LST radiator covers the pipework as well as the radiator, so that all hot surfaces are concealed. The routing and insulation of pipework also needs to be considered. Radiators in public areas should also be encased down to floor level so that younger learners, especially those who are crawling, cannot gain access to hot surfaces from below.

For spaces with high ceilings, such as sports and function halls underfloor heating may used. However, underfloor heating will not be suitable where large areas are covered with mats or where regular spillages occur.

Domestic hot water system

Educational buildings with younger learners must take measures to prevent scalding injuries from the domestic hot water system. This may lead to a design with hot water at two temperatures: a high temperature supply for kitchens, serveries and cleaners' sinks, and a low temperature supply for hand washing basins and younger learners' use. For the higher temperature, typically thermostats are set no lower than 60 °C. All hot water supplies that learners may have access to should have temperature reduced to a maximum of 43 °C by means of a thermostatic mixing valve (TMV) at the point of use. However, the storage and distribution temperatures cannot be too low, or there will be risks from legionella.

Ventilation system

There has been much research implying a positive relationship between indoor air quality in schools and learners' performance. Educational buildings are especially susceptible to poor environmental conditions: they contain numerous pollution sources, such as laboratory chemicals, cleaning supplies, chalk dust and mould, furnishing, volatile organic compounds and high levels of ozone, carbon dioxide and carbon monoxide all in areas with relatively high occupancy densities compared with, say, offices. Also, approaches to hygiene in younger learners may not be fully mature, meaning that surface and airborne infections may be more easily spread. This combination may affect cognitive function and reaction time of learners, and also cause medical issues such as eye, nose and throat irritation, headaches, loss of coordination and nausea.

Measuring and testing indoor air quality (IAQ) is an imperfect science with many variables. Measurement of the different element – such as carbon dioxide, carbon monoxide, radon, particles, volatile organic compounds and ozone – may be done with permanent monitors linked to alarms, or by regular measurements using portable equipment. IAQ can be improved by preventing outdoor pollution sources from entering classrooms as much as possible, and by providing adequate classroom ventilation to dilute the build-up of contaminants and subsequently remove them.

Sanitaryware

Size and fixing height of sanitaryware must be appropriate for relevant learner age groups.

Security systems

Security systems are required to protect learners, staff and the assets. There should be an overall security strategy. Depending on the type of establishment and the demographics of the learners, CCTV, access control, intruder alarms and asset tagging may be provided. A public address system may be beneficial in providing instant, effective communication to the whole school, particularly in emergency situations where a prearranged and rehearsed response to particular situations can be initiated, such as due to unwanted intrusions.

Some educational building may be particularly susceptible to arson attacks. This may be the driver for providing sprinklers.

Centralised clock and period bell systems

A centralised clock system may be provided to ensure that learners and staff are synchronised for classes and events, to avoid some learners roaming the halls while others are still in class. This may involve a hardwired system comprising a master clock linked to multiple slave clocks. The master clock may be a radio-controlled clock and have a radio receiver linking it to a time standard such as an atomic clock, or all the slave clocks may also be linked to an atomic clock.

Period bell systems may be used to denote the start of the daily school session and to identify the ends of various periods. The tones/bells should be easily distinguishable from the tones/bells used for raising the fire alarm. Zoned bell and tone distribution allows bells or sound tones to be rung in specific areas while not disturbing the rest of the facility, for example a five minute gym warning tone to sound only in the sports hall.

Special schools and those with designated units for learners with special needs must take account of learners' particular requirements. These might include:

- learners with mobility issues relying upon vertical transportation solutions to help move around a building; as well as lifts between storeys, this may include scissor platform lifts to transport a learner up a few steps
- particular requirements for sanitaryware and layouts in toilets and bathrooms
- learners with a hearing impairment needing higher light levels/clear visibility for lip-reading and signing and needs for being alerted to a fire alarm situation
- learners with a visual impairment needing higher light levels to facilitate way-finding to minimise the risk of accidents
- learners being very sensitive to glare from direct or reflected sunlight; for example, uplighters may be more suitable for some children with autism
- luminaires should be low glare, avoiding any flicker and unwanted noise; in particular there should be no disabling glare from light sources over changing beds or therapy couches
- automatic sensors that switch off lighting when no movement is detected may not be suitable for learners who are less mobile, but manual switches may be useful in teaching leaners how to use them
- learners with an inability to sense dangerous conditions, such as hot water or hot surfaces, may be inhibited or undeveloped or may have some other incapacity or inability to avoid such danger.

It is possible to monitor utility services consumption and generation in all or parts of buildings and display the information on an information dashboard. Education tools derived from the buildings performance can be used to educate learners, teachers and parents about building efficiency and to promote sustainability.

Index

References to figures are in *italics*; references to tables are in **bold**.

pre-construction, 168–71
pre-qualification questionnaire
    (PQQ), 44–5
prefabrication, 151
preference engineering, 109–10
pressure groups, 63
private finance initiative (PFI), 33
privatisation, 62
professional institutions, 63–4
professionalisation, 6–7
programme evaluation review
    technique (PERT), 212
project preparation, 157–8
public bodies, 51–2
public health, 26
public utility connections, 110–11

quality management, 220–1
quantity surveyors, 56–7, 218–19

radiators, 94–5
radon, 85
rainfall, 78
reasonable skill and care, 40
reference data, 166
reflected ceiling plans, 194
relationship capital, 21–2
reputation risks, 199
residual costs, 217
review process, 221
risers, 162–3
risk assessments, 173, 176
risk management, 198–202, **201**
    evaluation and quantification, **201**
    identification, 199–202
    sharing and management, 201–2
room data sheets, 89

safety, electric shock, 131
sanitaryware, 192, 249
security engineers, 59
security systems, 143–5
    alarms, 144–5
    closed circuit television (CCTV),
        114
    data centres, 243–4, 244
    door locks, 192
    hospitals, 238–9
    lighting, 143
    patrol stations, 145

sensors, 135
service scope, 14
servicing, 3
shop drawings, 175
shopping centres, 244–5
site acceptance tests, 178
site visits, 186–7
skills, 18
smoke detectors, 194, 232
snow, 78–9
soft skills, 18
software, 19–20, 196, 221
solar radiation, 77–8
soot, 85
space planning, 159–60
    heating, 163
    spare capacity, 164–5
    ventilation systems, 164
spare capacity, 164–5
spare parts, 167
sports facilities, 245–6
sprinklers, 142
stack effect, *115*
stairwell pressurisation, 93
stakeholder analysis, 185–6
stakeholder interfaces, *47*, 213
standards, 185
statutory requirements, 28–9, 50
steam systems, 236, 239
structural capital, 19–20
structural engineers, 56
structured wiring systems, 145
subcontracting, 19, 169
sulphur, 84–5
suppliers lists, 169
surveys, 33–4
sustainable urban drainage systems
    (SUDS), 78
swimming pools, 246
switchgear, 195

taps, 192
telephones, 149–50
temperature, 75–7
    kitchens, 230
    outdoor spaces, 104
    thermal comfort, 90, 94–5
tenders, 168–71
    documentation, 170
    evaluation, 170–1